# PERFECT FIGURES

OTHER BOOKS BY BUNNY CRUMPACKER

*Nonfiction*

The Sex Life of Food
The Old-Time Brand-Name Cookbook
Old-Time Brand-Name Desserts

*For Children*

Alexander's Pretending Day

*Cowritten with Chick Crumpacker*

Jazz Legends

# PERFECT FIGURES

## THE LORE OF NUMBERS AND HOW WE LEARNED TO COUNT

BUNNY CRUMPACKER

THOMAS DUNNE BOOKS
ST. MARTIN'S PRESS NEW YORK

THOMAS DUNNE BOOKS.
An imprint of St. Martin's Press.



www.thomasdunnebooks.com
www.stmartins.com

Book design by Gretchen Achilles

Library of Congress Cataloging-in-Publication Data

Crumpacker, Bunny.
Perfect figures : the lore of numbers and how we learned to count / Bunny Crumpacker.—1st ed.
p. cm.
ISBN-13: 978-0-312-36005-4
ISBN-10: 0-312-36005-3
1. Numeration—Miscellanea. 2. Counting—Miscellanea. I. Title.
QA141.C88 2007
513—dc22 2007016759

First Edition: August 2007

10 9 8 7 6 5 4 3 2 1

# CONTENTS

TO KARLA KUSKIN BELL
AND ALL OUR YEARS

*I have been to school most of the time, and could spell, and read, and write just a little, and could say the multiplication table up to six times seven is thirty-five, and I don't reckon I could ever get any further than that if I was to live forever. I don't take no stock in mathematics, anyway.*

**HUCK FINN,**
IN *THE ADVENTURES OF HUCKLEBERRY FINN* BY MARK TWAIN

*I'm very well acquainted, too, with matters mathematical,*
*I understand equations, both simple and quadratical.*
*About binomial theorem I'm teeming with a lot of news,*
*With many cheerful facts about the square of the hypotenuse.*

**"MODEL OF A MODERN MAJOR-GENERAL"**
FROM *THE PIRATES OF PENZANCE,* BY WILLIAM SCHWENCK GILBERT AND ARTHUR SEYMOUR SULLIVAN

# ACKNOWLEDGMENTS

Deepest thanks and appreciation to my editor, Peter Joseph, who is insightful, helpful, always of good cheer, and funny too; what a lovely combination! William Wolz read these pages in an early version, and made several helpful suggestions. I'm grateful to him, and to Cathy and Michael Wolz for helping me find him. Official thanks are overdue to Toni Plummer and Jess Korman—Toni for cheery and cool efficiency, and Jess for a host of helpful suggestions. I'm grateful for Jan Connor's eagle eyes, and to their happy owner as well, for Matt Hinton's discerning eye and mind, and to Lee Sennish, Sabine von Aulock, and Robbin Novak. And then there are special thanks for Chick, my own Ramanujan.

# PERFECT FIGURES

# 1

## ONE IS ALL

One is the beginning, the single starting place. It's the universe at the big bang: There was that enormous event, that unthinkable noise, and suddenly, in a fraction of a second, whatever there was—had it been One?—had shattered. It became a billion million stars, galaxies after galaxies of stars, stars with planets and moons and meteors and asteroids, each one containing everything again, atoms and molecules, charm and quark, and each thing—each atom, each galaxy—was still one, one again. One after one to infinity.

> Very unlike a divine man would he be, who is unable to count one, two, three, or to distinguish odd and even numbers. PLATO

At our own beginning, there were no numbers, not even one. We had no need for numbers—no need to count, no need to know how many. Sufficient unto itself, and for our survival, was each person, each thing, each moment.

We knew the day, and the darkness that came after the day, night after day and then day after night. But eventually the time must have come when someone wanted to keep track of yesterday, today, and tomorrow, and count what fills those days: moons and meals and springtimes. And eventually, someone wanted to count what was his—animals, perhaps, or arrows, or seeds, oil, and grain. Perhaps someone wanted to know what was coming—how many days it would be necessary to wait until the floods would come again, or the moon would disappear and then come back, how long before the baby would be born, or how long before the sun returned from its trip to the edge of the world and the cold passed and the days slowly began to grow longer again.

There is an unexpected story in the history of how we learned to count—from our first recognition of numero uno, the one we mean when we point to ourselves, to the vast numbers we think of when we look at the stars on a moonless night.

Counting is a natural process, almost inevitable, and numbers are organic. They begin with the single line of our bodies, the psychic feel of ourselves. They grow, one by one by one, but they always remain intimately connected to our physical being, from the twoness of our eyes and arms and legs to our ten fingers and—should we need them—our ten toes.

> The concept of number is the obvious distinction between beast and man. Thanks to number, the cry becomes a song, noise acquires rhythm, the spring is transformed into a dance, force becomes dynamic, and outlines, figures.
>
> JOSEPH DE MAISTRE

Counting is as natural as numbers. We count each other, and then our children, the things we own, the days we've passed through. Numbers count the things of the world, which must have become less untamable as we numbered its parts and learned to give them names. When we drew an animal on the wall of a cave, we made this one, and then that one—one and two—and we gave the animal its name. When we made designs of the stars over our heads and told their stories to each other, we remembered how many stars there are in Orion's belt or in Cassiopeia's chair. Everywhere, we learned numbers this naturally, and taught ourselves to count, because we always needed to know *how many*.

Once, we wanted to know how many cattle left in the morning and came home at night, how many seeds it was necessary to save for next year's planting, or how many days there were from full moon to full moon. Now we need zip codes and Social Security numbers, phone numbers and license plates. Our passports and our houses are numbered, as are our charge accounts and checks and telephones. Numbers have grown away from the simplicity of you and me and the baby. Now, they define us with increasing complexity, and we lose track of how they began, one by one by one—this one, that one, those, and me, you, and the rest of the world.

In another way, perhaps God taught Adam to count when he crafted Eve from Adam's rib and suddenly, where there had been one, lo! there were two. Did Adam and Eve keep track of their children by counting? Did they subtract one from two when Abel disappeared and only Cain was left? Apparently not. But clearly, God could count.

Once we left the garden, learning to count for ourselves was a slow process—it took millennia to learn to answer, in all the varied ways, the basic questions of mathematics: *how many? how much?* Like magic, numbers became visible as we needed to know them. Along the way, people counted in different ways in different places—by twos, fours, fives, twelves, twenties, sixties, and finally, by tens. Everywhere, numbers became part of civilization, and then came to mean more than just the amount they stood for—they meant good luck and bad, wishes and fairy tales, religion and a way of forecasting the future.

Although we learned numbers at different times and places, we always began with one. Here, we used piles of pebbles as equivalents for numbers, and there, we made notches on sticks or bones, but everywhere, even today, men and women have used their fingers to count. We forget how organic numbers are—how they grew out of our bodies as we learned to hold one finger up to mean one, or point to our eyes for two, and to find higher numbers on our fingers and our toes.

It isn't necessary to count the way we do now in order to answer the basic question of *how many.* Little children can say numbers from one to eleventy and beyond, but they have different ways of telling you how many toys they have—this one and that one and the other one, maybe, or just twenty-teen-two.

## COUNTING WITHOUT NUMBERS

Do animals count the same way we do? No matter how well we think of ourselves, we aren't alone in grasping the concept of number. Even the lowly rat can learn to press a lever a specific number of times, or take the third right turn (not the second or fourth) in a maze that leads to a reward. Pigeons can learn to peck at a target a specific number of times—thirty-five, not forty, or twenty-three, not nineteen. Chimpanzees will choose a tray that holds seven chocolates over one that holds six. Clever Hans, the famous counting horse, used almost subliminal cues from his trainer and figured out from that—not from the number—how many times to stamp his hoof, but all these animals are, in some way, counting.

Recent research leads us to believe that counting may be simply hard-wired into our brains, whether we're pigeons or people. Monkeys in a study at the Massachusetts Institute of Technology linked computer frames containing a number of dots to frames that were different but contained the same number of dots. Neurons in the monkeys' prefrontal cortex (the part of the brain that makes rapid decisions) rewired themselves when the first frames were shown, and when the same numbers reappeared, the neurons were quickly reactivated—the monkeys recognized the number, even though the frames were different.

But how do cicadas keep track of the years they need underground? They rise through the green grass and into the trees once every seventeen years to make a new generation, and then return through pale roots to the dark underground, to wait and count through all those slow winters before the precise springtime comes and they are ready to rise again, thousands of them all at once, to find their mates.

How do ants find their way home? Or, for that matter, how do they know the way to return to a food source, once they've found it? Researchers trained desert ants to travel from their nest to a food source, and then, to test whether the number of steps they had to take affected their ability to locate the food, glued stiltlike extensions to the legs of some ants and shortened (by cutting) the legs of others. The newly tall ants took the same number of steps they had already learned, and went right past their food. The shortened ants traveled only part of the way to their food goal. According to a 2006 issue of *Science,* as the ants became used to their new leg length, they adjusted their internal pedometers, in there counting away, and learned how many steps they now needed to take from the nest to the food and back again.

A group of wasps places dead bugs in the cells of developing wasps to be used as food. Different kinds of wasps use different numbers of bugs, but each is consistent, always using five, or ten, or twenty-four. In one group, ten bugs are always placed with female eggs but only five with males, because the female wasps are larger than the males. Somebody is counting.

Perhaps the bugs were arranged in patterns, to be recognized in the same way that we recognize the number of dots on a pair of dice. We don't actually count one-two-three-four, up to six; we look at the pattern and we know. Maybe that's what the wasps do.

Dolphins can recognize strings of as many as eight abstract figures. People, smart as we are, can go up to only six, or at the most, seven. We have to put hyphens in our telephone numbers—one after the area code, and another after what used to be called the exchange—because it's easier to remember a group of three numbers, another group of three, and a group of four, than it is to remember an uninterrupted group of ten. Zip codes stop at five; the added specific code of four more numbers is separated from the zip numbers by a hyphen. Is it counting to remember a phone number? Or is it the memorization of a group of words that happen to mean a number?

We don't usually stop and count anything we're looking at until the group goes higher than four or five. We look at the pattern—a square with something at every corner for four, and with something in the middle for five, or the shape of a triangle with something at each of the angles—and we know the number. We're recognizing the pattern, and we know the pattern means a number. A two-car garage has two doors and holds two cars, and that's as much as most of us need to know about it. But six and seven and eight and, of course, beyond—that's too many to count at a glance. We're not dolphins.

There was a group of Indians in South America without any number words beyond three. But they could recognize groups of things, and knew whether they were complete or not. When they traveled, they were accompanied by their dogs—and while they didn't know how many dogs they had, they knew at a glance if one was missing, and they'd call until the dog returned. Teachers on a class trip work in somewhat the same way. They have a feel for the way the group should look, and if it doesn't look right, *then* they stop and count. That kind of glance—the pile is smaller than it should be—gives you important information (something is missing), but it doesn't give you a number, it doesn't say how many you have. This kind of counting, the pattern at a glance (there's actually a word for it: *subitizing*), works only with relatively small numbers. If the teacher has thirty or more kids with her (and no parents, alas), she'll be hard put to know without counting when one is missing.

Up to a point, birds can do the same thing. Take one egg from a nest of several and the bird pays no attention. Take two, and the nest is abandoned. A nineteenth-century astronomer and mathematician, Sir John Lubbock, wrote about a landowner who was bothered by a crow that had chosen to nest in his watchtower. He'd go into the tower to shoo away the bird, and the bird would fly outside and wait until the man had left—and then it would fly back to the nest. After a while, the landowner thought of a way to fool the bird. Two men were sent to the tower with instructions for one to wait inside while the other left; the expectation was that when the bird saw a man leaving, it would feel safe and fly back inside, where the remaining man could deal with it. But the bird was smart enough to wait until they had both left before it returned to the nest. The next day, three men went into the tower and two men left. The clever bird wasn't fooled; it waited patiently again until all three men had left, and then flew happily back inside the tower. Four men entered the tower; three left; the bird waited for the last man before flying back home. At last, five men entered the tower; four left; and this time the bird flew back in again while the fifth man was still there. Conclusion: the bird could count to four, but not to five.

> The first principle is that you must not fool yourself—and you are the easiest person to fool.
>
> RICHARD FEYNMAN, COMMENCEMENT ADDRESS AT CALTECH, 1974

The bird was aware of quantity, even if not of number, even if up to only four. This is the beginning of counting. To be aware of the self is the first step toward being aware of the other; and knowing that there are two—me and you—is the first step toward counting the rest of the world. Eventually, we'll know that counting has no end—that it goes on forever, and even then, there can still be one more. In order to do that, to count to forever, we'll have to have the numbers that will make it possible to do so.

We can reach all our numbers, to count to forever, using very few numerals and words. We only have ten digits to work with, no matter what we're counting: 1, 2, 3, 4, 5, 6, 7, 8, 9, and 0. Each of those numerals has its own name—one, two, three, and so on, past ten to twelve. After that, for a long time number names are just a combination of the words that have gone before: thirteen is three and ten; twenty is two tens. Ninety-nine is nine tens and nine. When we add one more, we reach a hundred, the first new number name since twelve. The next is thousand, and the next after that—a long way away—is million—so few words and digits for so many numbers!

In exactly the same way, there are only twenty-six letters in the English alphabet, enough to encompass both Shakespeare and the Marx Brothers and everything before and after. A dictionary, said Anatole France, is the universe in alphabetical order. There was only one Shakespeare; there were three principal Marx Brothers; there are millions of words in the dictionary and numbers in the universe. But there are still only twenty-six letters in the English alphabet and a mere ten digits. Those ten numerals are enough to count all the words ever uttered by anybody, and all the letters in all the words. Further, the ten digits are all we need to count every grain of sand, every star, every *thing*—even every *no thing,* as the distance of empty space—in the universe.

> **It's not the voting that's democracy, it's the counting.**
>
> **TOM STOPPARD,** *JUMPERS*

## ONE WAS FIRST

The first number to be thought of, inevitably—because it was the first to be needed—was one, that inward number. One might be enough for a while: one is today, one is now, one is self, one is me, one is first.

As long as one is the only number you need, life is simple. If you have only one of anything, you needn't count it. It's either there or it isn't. You know right away: you see it or you don't. It's not until you add another and another—and another—that keeping track becomes complicated and you need more numbers. Numbers are a sign of plenty.

Even with plenty, though, you can get by with using just one for a while. The notches on ancient bones, the first written countings we know about, are simply lines of one, like notches on a gun stock or scratches made on a jailhouse wall when the sun goes down.

Those lines carefully etched into an animal bone must have come after the very first countings, the ones that we made on our fingers. The index finger points when it's held horizontally; it counts when it's upright. One finger; one number. Most counting systems are based either on five (one hand), ten (two hands), or twenty (fingers and toes). The Latin word for fingers is *digiti;* our fingers are digits, and numbers are digits as well. In the Middle Ages, *digiti* was used to mean the single numbers of the decimal system, while *articuli* (joints) meant the tens; then *digiti* came to mean the numerals themselves. When we talk about digital computing, we mean computing by numbers—rather quickly.

There are great advantages to counting with your fingers. With any luck, they're always there. They're clearly visible—other people can see them if you want them to. You can feel what you count. They're portable; wherever you go, they go with you. The problem is, though, that they aren't permanent. You hold up two fingers, and everybody understands that you mean two. When you put your fingers down, two disappears. It leaves no record. You can't walk around for days with two fingers in the air—you'll need them for other things. Unless you have more hands than most of us do, you can't count very high and you can count only one kind of thing at a time—birds or bananas, but not both. (More on finger counting, which can be surprisingly complex, in later chapters.)

There had to be better ways. And there were. Everywhere, always, there was someone who found a better way.

## ONE TO ONE

As human beings, we got our start in warm places—Africa, and what we call the Middle East. Fruit and grains grew naturally in our garden, and for a long time we were happy there, nibbling on the things that grew on trees. The years—the aeons—went by while we ate, until we were full of fruit, and so we were fruitful, and we multiplied. We began to wander, and in the new places we found, we hunted and scavenged and made our way, until gradually and slowly, we learned to tame the wildness—and that, in turn, is what tamed us.

We learned to plant seeds, and we stayed to harvest what grew from them. We had farms, and soon, we had neighbors. There were even villages—small groups of farms and people supplying each other's needs. If one family had cows, another might have extra grain, and a third olives for making oil, or grapes for wine. And thus we began to learn things beyond sowing and reaping and tending the animals, because we wanted to trade, and how can you do that—or sell—if you can't count? How can you measure your land if you don't have any kind of numbers? And if you begin to know ways to count—without knowing any numbers—how can you keep track of your countings?

After fingers, the first counting was simple: just one for one. When the cows went out to the field in the morning, the farmer made a pile of pebbles: one pebble for each cow. (Our words *calculate* and *calculus* come from the Latin *calculus,* a pebble, and the Greek *khaliks,* rock, or limestone.) At the end of the day, the cows came back, and one by one, the farmer removed the pebbles from the pile until all the pebbles were gone. If one pebble was left, one cow was missing. If somebody asked how many cows the farmer had, he could point at the pile of pebbles and say with authority, "That many." We look at an auditorium full of empty seats; the auditorium holds five hundred people. When the seats are filled, we know without counting how many people are there. Prayer beads, used by Muslims, Buddhists, and Catholics, are another kind of one-to-one counting: one bead for each prayer, so it's possible to pray without losing track of how many prayers have been said in the sequence.

The pebbles worked for a beginning, but they were only a bit more useful than fingers: They're slow going, one after one after one; piles of pebbles can tumble over; one pebble looks like another; and it's difficult to keep a permanent record. Anybody could accidentally knock the heap over, pebbles can easily get lost, and a pile of pebbles is hard to move around. Again, something better was needed.

What came next was remarkably permanent: notches in bones. The oldest—a baboon's thigh bone with twenty-nine notches—is thirty-five thousand years old. It was discovered in the Lebombo Mountains of Africa. A wolf bone, found in Europe, is thirty thousand years old. It has fifty-five notches. Our word *tally* comes from the Latin *talea,* cut, as a cut twig.

The bones (and wooden sticks) were easier to use than pebbles—a stick could be held in one hand, and as each thing, each sheep, perhaps, came back to the fold, the thumb could slide from one notch to the next. When a new lamb arrived in the springtime, a new notch could be carved to represent it. The sticks could be carried—and they could be split in half, so that a borrower could have a permanent and foolproof record of what he had borrowed. When he made good on his debt, the two halves of the notched stick could be matched—and the record of exactly how much had been borrowed would be clear.

Notched sticks are one of man's first inventions. They came

after discovering how to use tools for hunting, but before inventing the wheel. And so many millennia later, we still count this way when we need to. Men working in Southern California used to keep track of their days by cutting a line for each day in a block of wood; they used a deeper or thicker line at the end of each week, and a cross at the end of each fortnight. Cowboys made notches in their gun stocks for each buffalo they killed—or for each Indian. During both world wars pilots drew decals on their planes to mark the number of enemy planes they had downed, another kind of tallied one-to-one record. In England, notches on wooden batons stood for varying amounts of pounds sterling. Charles Dickens, in a speech given in 1855, told about

> a savage mode of keeping accounts on notched sticks . . . introduced into the Court of Exchequer, [where] the accounts were kept, much as Robinson Crusoe kept his calendar on the desert island. . . . Official routine inclined to these notched sticks, as if they were pillars of the constitution, and still the Exchequer accounts continued to be kept on certain splints of elm wood called "tallies." In the reign of George III an inquiry was made by some revolutionary spirit, whether pens, ink, and paper, slates and pencils, being in existence, this obstinate adherence to an obsolete custom ought to be continued, and whether a change ought not to be effected. All the red tape in the country grew redder at the bare mention of this bold and original conception and it took till 1826 to get these sticks abolished.

The English Royal Treasury had kept its accounts on tally sticks since the twelfth century. Income, expenditures, taxes—all were notched on tallies. And more: double-notched sticks were issued; half the stick, with one set of notches, was redeemable (assuming it matched—or tallied) for cash through the Treasury, which held the matching piece. Thus, the written (or notched) certificate payable to the bearer after it had agreed with its security: a check. The words *stock* and *dividend* also trace back to the Treasury's tally sticks—the stock was the specially marked tally stick held by anyone who lent money to the Bank of England, and thus became a share—or stock—holder. A dividend originally was a *tallia dividenda*—a "stick to be divided"—redeemable through the Treasury.

The Treasury's Court of the Exchequer didn't take its name from bank checks. Instead, it was named after the cloth that covered the table in the room where local administrators came to settle accounts with the Crown. The cloth was checkered so that counters could be placed in squares that corresponded to various amounts; the final sum was entered both in an account book and on a tally stick—and everyone could understand the process, even if they didn't know how to read or write.

In 1782, it was decided that tally sticks would no longer be issued by the Royal Treasury, but they remained valid until 1826. In 1834, the process of burning the vast numbers of outdated tally sticks began in the furnaces beneath the Houses of Parliament. Unfortunately, the fire was so intense that the Parliament buildings themselves caught fire and went up in flames. In his 1855 speech, Dickens continued his description of the Court of the Exchequer's use of tallies to keep accounts. "It came to pass that [the tallies] were burnt in a stove in the House of Lords. The stove, overgorged with these preposterous sticks, set fire to the panelling; the panelling set fire to the House of Lords; the House of Lords set fire to the House of Commons; the two houses were reduced to ashes; architects were called in to build others; we are now in the second million of the cost thereof; the national pig is not nearly over the stile yet; and the little old woman, Britannia, hasn't got home to-night."

Britain wasn't alone in its use of notched sticks. Tally sticks were used in Germany, in Switzerland, throughout Scandinavia, in Indochina, and in a host of other places. The word for *contract* in Chinese is written with a character made up of three parts: the character for a tally stick on the left, knife on the right, and under both, another character that means large. Thus, a contract is a big tally stick.

In *The Universal History of Numbers,* Georges Ifrah writes about a French bakery where, until as recently as the 1970s, tally sticks were used as credit cards of a sort. Two small pieces of wood, called *tailles,* were notched every time a customer bought a loaf of bread; the baker kept one plank, and the customer took the other home with the bread. At the end of the week, the two planks were matched—the notches had to be equal—and the bill was settled.

## AHA!

Tally sticks lasted because they work. But even so, even though they're portable and permanent, they have drawbacks. They probably began before settled farming; but once groups of farms had turned into towns and villages, something more, something better, was needed. Land had to be divided and borders established. Workers had to be paid with supplies of grain or jars of oil. Crops had to be kept track of, so that there would be food to last through the seasons, with enough seed reserved to plant again in the spring. Animals had to be counted and recounted. Taxes and tributes had to be paid. Days had to be numbered so that farmers would know when the moon would be full and thus the days longer for working in the fields, and when the sun would fade in the southern sky as the days grew shorter and the nights cold and long.

Carving into bones was cumbersome; wooden sticks were relatively fragile and flammable. Farmers began instead to fashion clay tokens to serve the same purpose—one-to-one counting, still done without numbers. Tokens were easily moved from one place to another, and they lasted as long as was needed. Most important, they made the next step more obvious: there could be more than a single shape. A round token could be equivalent to one of whatever was being counted, and an oval shape, ten. And then: how much easier to have ten oval tokens instead of a hundred round ones! What a simple but amazing leap forward—no less so because from our point of view it may seem inevitable. The road is rarely clear without a map, and the first counters were traveling in an unknown country of numbers.

There were many moments of genius along the way to where we are now (and there are undoubtedly more to come)—pure Aha! moments that seem so simple when we look back at them, but were so complicated and hidden before somebody thought of them. We learned to make tokens instead of carving bones, and we learned to make simple shapes that represented different amounts. The next leap was considerably larger: tokens could be shaped differently not only to show quantity, but also to show the object being counted—jars of beer or wine could have one shape; loaves of bread another; you could tell by looking (without written words) what it was that was being counted. First the object-shape, and then the number-shape.

> No man acquires property without acquiring with it a little arithmetic also.
>
> RALPH WALDO EMERSON, *REPRESENTATIVE MEN*

It was the Sumerians who took this idea and made it marvelously bold. They lived in the Fertile Crescent, the sweet soil of the land between the Tigris and the Euphrates rivers, in what is now southern Iraq. Their early settlements grew into towns and villages, and then into cities. The largest was Ur; about twenty-five thousand people lived there, with an additional twenty thousand in what we'd now call the metropolitan area. There were other large cities in the Fertile

Crescent, and inevitably they began to specialize and to trade—timber for gold or silver, barley and corn for oil and wine. In order to trade, they had to ship their products by land or by water, and they had to keep track of what was going and what was coming.

By the second millennium BC, they had the clay tokens. But problems remained. Someone might send twenty-five weights of grain, which would be accompanied by twenty-five tokens. There was no way to stop the shipper, if he was greedy and daring enough, from taking one weight and one token—the receiver would never know what had happened, that he wasn't receiving what he was paying for. The seller, equally unknowing, could also be short-changed. In order to prevent all this from happening—and keep suspicion at bay—the sender had to know exactly what had been sent; the record had to accompany the goods while they traveled; and the receiver had to know he was getting what he'd paid for.

Clay tokens couldn't do the job—too easy just to take one. They could be put in a pouch—maybe that would work—but then, anybody could open it. Something that would last was needed, to accompany the shipment until it had arrived safely. The solution? Wrap the tokens in clay, and seal the clay-wrapped tokens by baking the whole thing. When the shipment arrived, the buyer could break open the clay package and count the tokens inside. A match made everybody happy.

For a while. Once the packages were broken open, their usefulness was over. All that was left was a pile of tokens. Given that there were three parties involved in each transaction, the sender, the shipper, and the recipient, each had to know that the other two were honest—that the count matched the shipment from beginning to end. There had to be another step.

What if the sender drew a picture on the *outside* of the clay wrapper showing the number of tokens inside? Then the package could be sealed so that if anybody tried to open it, the receiver would know. More: what if the sender used the same theory that the tokens themselves were based on? He could draw one shape to represent grain, or another to stand for oil, and then he could make a second set of symbols to show quantity—one shape, a simple line, to mean the number one, another for ten. Now there would be a picture on the outside that *exactly* matched the inside, with the sender's seal to validate the package. That would do it! When the seal was broken and the bag was opened, the tokens inside would match the picture on the outside, and the shipment would be safe.

But Aha! After all that, having come this far, why bother with the inside, with all those pesky little tokens? Everything that everybody needed to know was clearly depicted on the *outside* of the clay envelope. Forget the tokens! When the shipment arrived, the shipper could double-check to see that the figures on the tablet matched the shipment, and the purchaser could clearly and easily tell whether or not he'd been cheated, or if was dealing with honest men. The wrapper was no longer a wrapper. *It was the thing itself*. It was a clay tablet. *It was an invoice*. And the pictures were *written* on it. It was a count, it was a record, and it was writing. *Counting came before writing*. Counting *was* writing. We counted because we needed to, and then we learned to write down what we had counted. And from there, we flew down the years to poetry and libraries and shopping lists and constitutions. One, two, and three. And more to come.

## ANOTHER NOTCH

For such an important number, the beginning of everything, one is quite plain. It looks simple. It has no subtlety; it just stands there, straight and tall, entire and unadorned. It has no curls or loops, no corners to speak of, nothing soft, no womanly curves, just one firm line, top to bottom, solid. At the most, it has half an arrow on the top—or, if you prefer, a little hat, tilted to one side

> One is one
> and all alone
> and ever more
> shall be so.
>
> "GREEN GROW THE RUSHES,"
> ENGLISH FOLK SONG

in a somewhat jaunty attempt to look debonair. It sometimes has a tiny platform on its bottom to stand upon, but it remains modest and single, somewhat stubborn, unbreakable as it is except into fractions. Its very solidity and stubbornness make it quite safe, but there's no connection there. I think of one as lonely and a bit dull—or at least rather bored. Alone and lonely, after all, both have one in them; and only is just one, and none is worse, not even one.

One can indeed accomplish so much (one plus one forever), but it's left with so little. It must be filled with longing (it's a bachelor—definitely phallic), yearning for another, as Adam did, all alone in Eden. There are solitary joys, and some of them are splendid, but so much of ourselves is built around the idea of two: eyes, ears, hands—two is almost as basic to our natures as one. More so, if the need to share is recognized. "Look!" we want to say when we see something unusual. Or "Did you hear that?" One is the beginning, the essence, one is what everything is made of, and what we must always return to, but there's no getting around it—one is alone, and can be lonely. One is myself, alone at first and alone at last.

But whatever comes next, one was here first. And in the other direction, one is all that separates us from nothing, the void, extinction—zero, though zero arrived a long time after one had left nothing behind. One was first.

One is distinction and difference. One is apart; one is known. One is a place, a thing, a dot, a point. There is no counting without one, though one doesn't need to be counted—it's simply there. But then, there is no anything without one.

One is the *integer,* the whole number, and one is integrity. The simplicity and singularity of one are deceptive, for one is everything. It embraces all, and encompasses infinity. It can stand a little loneliness.

The first number word in English, *one,* comes from the Latin word *unus. Unus* is the root for words like *union, unity, unison, unique* (one of a kind), and *unanimous* (of one mind). It's also the source, indirectly, through Anglo-Saxon, for *a* and *an,* both of which mean one. *An apple* is one apple; *a banana* is one banana. To be all of one piece is to be whole, which gives us words like *holy, wholesome*—even *hale* and *healthy. Atone* is one too, to be at one again, and so is *once,* when it happened first.

The earlier Latin for one was *oinos*—strangely like an onion (which could be defined as one sharply tasty sphere encompassing many layers). The Indo-European prototype was, variably, *oi-no, oi-ko,* or *oi-wo.* In Sanskrit, the word is *eka;* in German, *ein;* in Russian, *odin;* in Irish, *oin.*

Old Egyptian hieroglyphs through all the kingdoms and dynasties show the number one as the familiar straight line. Egyptian carvers, in the later years, apparently liked puzzles and word games, and in some of their inscriptions used pictures to show numbers. One was sometimes depicted as a round circle—the sun, because there's only one sun, even if it does keep reappearing—or it could be shown as a small upward curved line with the top of a circle peeping over the top, representing the moon, for the same reason.

In something of the same spirit—though not as a puzzle—Indian mathematicians used a variety of poetic concepts to express numbers. The numerals we know trace back to ancient India; in the eleventh century, a Persian astronomer wrote about the numbers used in Indian astronomical tables, and said that when it was difficult to write the word for a number in a certain place in the tables, astronomers could choose from "amongst its sisters." Quoting the great seventh-century Indian astronomer and mathematician Brahmagupta, he went on, "If you want to write one, express it through a word which denotes something unique, like the Earth or the Moon." Among the other sister words for the number one were "the Ancestor," referring to Brahma, considered the creator of the universe, "Beginning," and "Body."

## THE ROMANS AND THE GREEKS, FOR A CHANGE

The Latin words printed on American coins tell us that one means coming together in unison: *e pluribus unum*—out of many, one. The ancient Greeks—the Pythagoreans—saw one differently, in the opposite way: out of one, many.

Numbers have always been a way of explaining the world, whether in Sumerian tokens and tablets, Einsteinian formulas, or the all-encompassing view of the Pythagoreans. Pythagoras was born in about 580 BC on the island of Samos, traveled as a youth to Egypt and Babylon, and eventually founded a school in Magna Graecia (southern Italy) which became the basis of the secretive Pythagorean Brotherhood.

Three hundred rich and powerful young men attended the school. They were divided into two groups: the outer circle, the *akousmatikoi* (those who hear—related to our word acoustics) learned the group's rules of conduct. Once these were mastered, its members could progress to the inner circle, called the *mathematikoi*. (Pythagoras is credited with having coined the word *mathematics,* to mean "that which is learned," as well as the word *philosophy,* "the love of wisdom.") They studied the most secret and difficult of Pythagoras's truths—in the areas of geometry, astronomy, number theory, and music.

Music was basic to their studies, for Pythagoras believed that the universe sings—that there is a literal music in the skies, formed by the motions of the planets and made up of tones based on numerical frequencies and ratios. He believed numbers enable us to understand that music's harmonies—but even beyond that, he believed numbers existed *before* there was physical reality, even that numbers create and *are* physical reality, the building blocks of the universe, the cause of everything that exists. Only through number and form, the Pythagoreans held, could man grasp the nature of the universe. Everything that *is* can be numbered.

> The odd and the even are elements of number, and of these the one is infinite and the other finite, and unity is the product of both of them, for it is both odd and even, and number arises from unity, and the whole heaven, as has been said, is number.
>
> ARISTOTLE

All of this began when Pythagoras discovered the relationship between the length of a string and the sound it makes when it's plucked; he realized that if the string is shortened to half of its original length, the tone it makes is an octave higher than the original note, a sound he described as pleasant and "in harmony." He then worked out the string lengths necessary for other harmonies—at three-quarters of its original length, the string produces a tone which we call a fourth; at two-thirds of its length, the tone is a fifth. While we know that the tones change because the vibrations of the string change at its different lengths, Pythagoras believed instead that the harmonies depended on the ratios between numbers, and that, eventually, all things depend upon numbers. For him, the heavens were a musical scale.

> Music is the arithmetic of sounds as optics is the geometry of light.
>
> CLAUDE DEBUSSY
>
> I myself figured out the peculiar form of mathematics and harmonies that was strange to all the world but me.
>
> JELLY ROLL MORTON

(In a way, Einstein believed something similar. He once said that Beethoven created the music he wrote, but Mozart's music "was so pure that it seemed to have been ever-present in the universe, waiting to be discovered by the master." Einstein saw physics in that way—that beyond observation and theory was the music of the spheres, a "pre-established harmony," waiting to be discovered.)

Pythagoras believed in the migration of the soul after death and developed a ceremony to purify the soul in readiness for its journeys. Because the souls of friends might return as animals, vegetarianism was preferred to meat eating. But the eating of beans was banned. Aristotle said that among the reasons for the bean taboo was the possibility that beans may have arisen simultaneously with humans in the moment of the universe's creation—and also that beans resemble genitals.

The Brotherhood was religious as well as mystical, and practical as well as either. It developed principles that influenced Plato and Aristotle, and much of Western rational philosophy. The first mathematical proofs were the Pythagoreans'; they developed mathematical theory still in use today—you learned about right triangles in high school geometry. (The sum of the squares of the lengths of the two shorter sides of a right triangle is equal to the square of the hypotenuse, which is the longest side.)

But beyond the hypotenuse, much of the Pythagorean tradition is in the realm of mystical wisdom, rather than science and scholarship, mathematics and geometry, as we know them today. A good example is the Pythagorean belief that odd numbers (one, three, five, seven, nine . . .) are indissoluble, therefore masculine and celestial in nature, and that even numbers (two, four, six, eight, ten . . .) are soluble, therefore feminine, ephemeral, and earthly. This male-female duality existed for the Pythagoreans beyond the male-female division of numbers. They divided reality into two parts: the mind and the spirit were aligned with the realm of the gods, which was male; the body and matter were the realm of the earth, which was female. Overall, numbers belonged with the male; thinking about numbers and the tasks of mathematics were masculine work, associated with the gods and with transcendence from all that was material. "It is here," Margaret Wertheim wrote in *The New York Times* in 2006, "that we begin to see the seeds of modern physics." The first universities were founded to educate the clergy. Women couldn't be priests, so neither could they be university students, and physics departments, well into the twentieth century, were often the last to admit women as students or as professors.

(Meanwhile, on the other side of the world, the Chinese believed that odd numbers meant white, day, heat, sun, and fire, and even numbers the opposite: black, night, cold, water, earth. The Chinese yin and yang represent odd and even numbers. The yang is masculine, and represents the sun, day, summer, light, and openess. The yin is feminine—the moon, night, winter, shade, secrecy. They alternate, one after the other, together and touching, one and two forever.)

For the Pythagoreans, individual numbers not only had abstract qualities like masculinity and earthliness, but also could be identified with human attributes. One, because it's unchangeable, they linked to reason; two, with its pairs and opposites, to opinion or polarity. Harmony was a property of three. Four stood for justice because it's the product of equals (two plus two) and because squares, with their four even sides, are perfect figures; four also represented space and matter. Five represented marriage because it's the first union of a feminine number with a masculine number (two plus three equals five)—and so on through the numbers to a perfect Pythagorean ten.

The trouble with one, for the Pythagoreans, was that for all its unchangeable reasonability, they didn't believe one to be a number at all. Numbers were totalities composed of separate units—something plus something—but one, they thought, is a totality complete unto itself. (They had no truck with fractions.) For them, there was one, and there was more than one. One is "that which is"; it's a statement of existence: one *is*. One doesn't have separate units; it *is* a unit, *the* unity, the opposite of the idea of many, the opposite of plural. One is pure and strong, and they believed it to be the single indivisible thing from which numbers arose—not the *other* numbers, because one was not a number, but *the* numbers. Numbers were an idea sort of like the offspring of the first man, the one, who was Adam. One is the father number. (And two—like Eve—is the mother number. But we're not at two yet.)

That concept of one, as the source of numbers without being a number itself, lasted through the Middle Ages. A twelfth-century manuscript, the *Salem Codex,* said, "Every number can be doubled and halved, except for unity; this can, it is true be doubled, but not halved—wherein lyeth concealed a great Mysterium [God]." A sixteenth-century German mathematician agreed: "1 is no number, but it is a generatrix, beginning, and foundation of all the other numbers." In *Number Words and Number Symbols,* Karl Menninger quotes Michael Stevin as the first mathematician to say—in 1585—that one is a number. He proved it by saying that if you subtract one from three, you are left with two; if one were not a number, subtracting one from three would leave three untouched.

Thousands of years after the ancient Greeks, Freud found one to be a perfect phallic symbol. It is upright and solid, as neat a phallic symbol as a stalk of asparagus—better, probably, because it isn't green. Whether it's because of its perfectly pricklike shape or whether it is some kind of numerical and linguistic testimony to the nature of the male, there are languages in which the word for *one* and the word for *man*—and more, even just the word for *penis*—are the same: man, penis, one—it all comes to the same thing.

There are also languages—other languages—in which the word for *God* is the same as the word for *one,* and *two* is the same as the word for *sin*—because two is the first step away from one, and therefore is a step away from God. In other languages, early Hebrew and Arabic among them, numbers began with two—as they did for the Pythagoreans, but here it was because one was the number for God (as "There is one God," or "God is One"), and was reserved for God alone.

## FIRST AND ONCE AND ALWAYS

Mathematically, one is the first odd number. But then, one is the first everything—one almost defines first. You can't have anything first without having one.

*First* is a word that does hard work as an adjective as well as a noun. It covers everything from the big toe on your foot to the opening lines of a poem to the base immediately to the right of home plate in baseball. It deals with time—it can be the earliest thing—as well as your car's lowest gear; it comes before all others in rank or occurrence; it's the beginning and it's the winning number. The president's wife is the First Lady; someday, perhaps the president's husband will be the First Gentleman, if not the First Man. On a first-name basis is the beginning of friendship. First place is best.

> **The last thing one knows when writing a book is what to put first.**
>
> **BLAISE PASCAL,** *PENSEES*

*First* is the ordinal word for *one,* which is a cardinal word. In other words, one is a number in a series—one, two, three, eventually telling you how many there are; first—or second or third—tells you the number's position in the series.

*First* and *second* are words that have nothing obvious to do with one and two, though clearly third, fourth, fifth—and so on—are derived from three, four, five. . . . The same thing is true in other languages—all the way back to Greek and Latin and before. *First* is related to words meaning "before," or "in front"—the Indo-European *pro,* "before," and the

Sanskirt *puras,* "in front," eventually evolved (with the *fr* sound substituting for the *pr* sound) into *foremost,* and then to *first.* (In German, *first* is *erst,* which originally meant "early morning.") *Second* is related to words meaning "the other," as the Latin *alter,* and the Indo-European root *anteros,* meaning "the other of two" or "the following," and the Latin *sequi,* "to follow," and *secundus,* "the next after the first."

In many places, counting went only from one to two—one was singular, two plural, and beyond two was a vast multitude, the idea of *many.* In the same way, it may be that it took us just that long to get the connection between the cardinal count of one and the ordinal place of first, and then two and second. Second was disconnected from two; rather, it was simply the next. By the time we reached three and third, we'd not only understood the number after two, we had also begun to understand the relationship; third wasn't just one after one after one. It was one in a series—the third, in fact. We're not that different from a tribe on Papua New Guinea. In their language, Ponam, the only words that show placement are first, middle, and last—nothing else matters, even though Ponam has many numbers to count with.

> The advantage of a bad memory is that one enjoys several times the same good things for the first time.
>
> FRIEDRICH NIETZSCHE

*Primary* and *secondary,* again from the Latin *primus* and *secundus,* "first" and "second," are followed by *tertiary* (third), *quaternary, quinary, senary, septenary, octonary, nonary, denary, duodenary*—which takes us to twelfth—and then there's a leap to *vigenary,* twentieth. There was a tiny leap from tenth to twelfth—the eleventh place has no single word of its own, alas. In the same way, *once, twice,* and *thrice* are followed by a total blank—there are no single words for four times, five times, or any other number of times; apparently, says the *Oxford English Dictionary,* a usually reliable source, our language has never needed any numerical words after *thrice*. Yet. Note that quince is a fruit, and has nothing to do with how many times you've been able to find one growing on a tree. Its name derives, in rather roundabout ways, through a variety of words basically meaning a kind of apple—though if you cut an apple in half across its middle, you'll find a five-pointed star.

> "Can you do addition?" the White Queen asked. "What's one and one and one and one and one and one and one and one and one and one?"
> "I don't know," said Alice, "I lost count."
>
> LEWIS CARROLL, *THROUGH THE LOOKING-GLASS*

Another quirk in this category of numerical word oddities is the pronunciation of *one*—and for that matter, *once*. Words derived from *one,* like *only, alone,* and *atone,* give the *o* a long sound—what Mrs. Quinn, my fifth-grade teacher, would have said was a letter saying its own name. *O* as in go, toe, row, over, and oh. Apparently, the way we pronounce *one*—as wun instead of w-oe-n—can be traced back to western England and Wales in the Middle Ages, when the vowel sound began a series of changes—from *o* as in own to *o* as in boot to *o* as in took and then *o* as in one. The final sound stuck and then it spread. Other vowel sounds changed too in those places (oats, for example, went from the long *o* sound of oh-ts to the sound of the *u* in cuts) but only the pronunciation of *one* and *once* stayed around to sound as we use it today.

The pronunciation of *one,* then, is an oddity, and one is an odd number, at least mathematically. Odd numbers can't be divided easily into two equal piles—there's always something left over. They're solid and smug. Like Popeye, they seem to say, "I yam what I yam." And they've always just eaten a can of spinach.

At heart, though, there's nothing odd about one. It *is* a number; it's a splendid number. It's part of the endless series that reaches to the stars but begins, always, with the number one.

If you have one number, you have every number, because every number has to begin with one, and because no matter what

> All for one,
> one for all.
>
> **ALEXANDRE DUMAS THE ELDER,** *THE THREE MUSKETEERS*

number you have, no matter how big a number it is, you can always make it larger by adding one more. Isn't that what infinity is? The last number and then one more . . . and one more . . . and one more . . . forever.

One is first because, just as the Pythagoreans said, all the other numbers came from one. Without one, there is nothing. With one, there is everything. Everything begins with one. One makes it possible to go on forever.

One is all.

# 2

## TWO IS YOU AND YOUR IMAGE IN THE MIRROR

In his poem "On Turning Ten," about the years in a child's life, Billy Collins writes about the perfect simplicity of one yielding to the beautiful complexity of two. Quite apart from the terrible twos, two is indeed full of contradictions and harmonies—opposites and pairs. Two is antonyms and homonyms, the beginning and the continuation, the point becoming the line. Two is the duality, the beyond—not yet the many, but no longer the one. Two is the beginning of magic.

Two is plural by its very nature, but there is a host of words that embody the idea of one as well as two: scissors, most obviously. It's a singular word, but it means two—as in a pair of scissors; it ends in *s,* and it has two parts. Then there are trousers, pants, eyeglasses, pliers, shears, tongs, jeans, knickers, pajamas, shorts, tights, binoculars, forceps . . . two and one, both. Deer and sheep go in the other direction, one as well as two, but still with the same idea.

> We are the two halves of a pair of scissors, when apart, Pecksniff, but together, we are something.
>
> CHARLES DICKENS, *THE OLD CURIOSITY SHOP*

There are languages, Hebrew for one, that have a special form of the plural for some pairs of identical things—hands, for instance, or feet—or for words naming things that have identical parts, like eyeglasses or scissors. In Hebrew, the normal plural ends in *im* for the masculine case and in *ot* for the feminine; but the plural for pairs or for things with identical parts ends in *ayim.* The ancient Greeks had separate words to show one thing, two things, and more than two things. One citizen was *polites;* two citizens was *polita;* and more than two, *politai.* That meant a singular, a dual, and a plurality, one, two, and many.

In some languages, a noun is repeated to indicate that there's more than one (as we say tra-la-la for the notes in a song, or boom-boom for a lot of noise, tweet tweet for what a bird says—or maybe even oom-pah-pah for the repeated bass notes of a tuba). There are also languages which don't have separate singular and plural, just one word that encompasses both.

There have been tribes that had no word for *number,* though they had numbers. For them, the *idea* of numbers, of counting, was not yet there. They might have words that meant one and two, but they were not numbers—*one-cow* (as a single word) might mean a cow, but it didn't mean one. In the same way, there were words for specific colors, but no word meaning just *color,* the overall idea of color. This kind of counting—without number—means that the idea of number, of two, for instance, isn't separate from the thing being counted. There's no abstract idea of two, of twoness. We remember this way of counting when we use phrases like "a brace of birds"—which can only be *two* birds; it can't be three, or four.

> **Students achieving oneness will move on to twoness.**
>
> WOODY ALLEN

"A brace of birds" is what's called a collective phrase—words that connect the idea of number to the name of a thing. Such phrases go way back, some as far as to the times when humans first learned to count, because we needed to express quantity before we arrived at numbers, and collective phrases were a kind of counting. A brace of birds is two birds; a yoke of oxen is two oxen. A couple or a pair are also two. They conveyed the *idea* of two before we had begun using the *number* two. There's a difference between "brace" and "pair," though. A brace of birds must be two *birds* and only birds, while a pair and a couple can be two of anything. A pair and a couple is that much more sophisticated—at least in terms of numbers. (More about collective phrases when we reach four.)

> **It must have required many ages to discover that a brace of pheasants and a couple of days were both instances of the number 2: the degree of abstraction involved is far from easy.**
>
> **BERTRAND RUSSELL**

## TWO BY TWO

Two is the next way to count after the series of notches that all mean one. It's as if one broke apart and became two, and then stopped there for a while. There's a bond between one and two—the other half of the pair, the man and woman, the woman and child. We ourselves are a study in two, our bodies a testament to its power: two arms, legs, heads, feet, eyes, ears, testicles, breasts—all two. Even our noses seem to be divided into two (The Indo-European root word *naso* means literally "two noses"—two nostrils combined in one nose.) In the sky above us, we see two, the sun and the moon; our days beneath are divided into two—the night and the day, the evening and the morning, the sunset and the dawn.

The first countings that used numbers, in many places, used only those first numbers, one and two. Two may have seemed like a magical number at first, with its physical manifestation in our bodies, and the first recognition of a single other as the humanity which is beyond ourselves. The Indo-European number word *duuo* shows not only in the Latin *duo,* but also in the German *du,* the French *tu,* and the English *thou.*

It may also have been that two was all the counting that was needed—beyond two was a vast immeasurable *many.* (There is more about *many* yet to come when we reach three and four.) In the different places where we learned numbers, we didn't learn to count them all at once, from one to thousands—we went single file, one by one, through the fingers on our hands, and two was what came after one. Karl Menninger, in *Number Words and Number Symbols,* calls two "early man's first hesitant step toward counting." We didn't go from two to ten to a thousand; two is where we stopped to catch our breath—it's where we landed after we had left the safety of one. "The number 2 is a frontier in counting, the first and oldest" of the many that there are along the way.

## COUNTING BY TWOS

Two-counting was widespread in Africa, South America, and Australia. The written sign for two evolved from the single line which was one. The first symbol for the number one was often a pebble, a single notch on a bone or a stick, and then a token that resembled a pebble or a bead. But it was almost always a finger, a single upraised finger. How many? One. One finger. So the written sign became a line that looked like a finger, sometimes vertical—or sometimes a horizontal line, as if it were a stick still lying on the ground. Two, then, would be two fingers, two vertical lines, or two horizontal. If you write the vertical lines quickly, they can look like N, and from this comes the vertical line with a curve on the top that means two in Arabic. The horizontal lines, written quickly, look like Z, and our 2 derives from the figure. The two unjoined horizontal lines are still used in China and Japan.

Counting by twos was widespread in Africa, South America, and Australia, but words for two were much less universal than symbols. In a South Seas island tribe, the word for one was *urapon;* for two, *ukasar.* To go higher, the words were combined, by twos. Three became *ukasar-urapon* ("two-one"); four, *ukasar-ukasar* ("two-two"); five, *ukasar-ukasar-urapon* ("two-two-one"); and six, *ukasar-ukasar-ukasar* ("two-two-two"). They counted higher than two in this way—without having new words—but how tiresome this kind of counting would become if the numbers went much higher: endless repetitions of *ukasar* and *urapon.*

(A variation of two-counting used three different words, for one, two, and three. Six, then, would be two threes instead of two-two-two, just that much easier.)

> **Ninety percent of this game is half mental.**
>
> **YOGI BERRA**

We're used to the repetitions of our own numbers: ten as *-teen* (thirteen, ten and three) and as *-ty* (twenty, two tens). But if we used only two words, like *ukasar* and *urapon,* over and over, it would eventually be not only tedious but also confusing.

Two-counting doesn't involve fractions, though it sounds as though it should. There wasn't, for example, a word which meant a fraction of *ukasar* or *urapon.* There was no concept for any word to match. There was either one or two; size didn't matter. When we divide something in half, we see two equal halves. When people who counted by twos divided things in half, they saw two things where before there had been one. Equal had nothing to do with it.

The idea of two also had nothing to do with it—we were able to count to two long before we understood the nature of two. The Tauade in New Guinea had different sets of words for two men and for two women, something like our "brace of birds." Other people used different counting words depending on what they were counting. One North American tribe used different words to count live things, round things, or long things.

The world was incredibly close and alive to the first two-counters, each thing within it separate and vivid. The idea of two came long after the number two; until the idea of two had been grasped, what mattered was not the number itself, but the thing being numbered. The word for two itself sometimes embodied a constant of the physical world—wings, eyes: these were two. Brahmagupta wrote that words that come in pairs—even opposites like black and white—could be used as the *name* of the numeral two.

English has many traces of counting by twos, even aside from *brace* and *yoke*—an amazing number of words mean two: a couple, a pair, a duo, a duet. . . . Twins are two children born of the same mother at the same time; the word can refer only to two. (Then there are triplets and quadruplets and quintuplets—who ever thought there would be a need for more?) Suspenders come in twos, to match shoulders; you can call them braces if you like.

One kind of counting by twos has lasted. Computers count by twos because their systems are based on electricity, and they can understand only two things: on and off. They have only two fingers, as it were. If we too didn't have fingers—just flippers, say, or wings—we might also have learned to count by twos, instead of the tens that match our fingers.

The two numbers computers use are one and zero—1 and 0—sufficient, given the power of their incredible speed, for computers to count from one to enormously high numbers.

> Two is not twice one; two is two thousand times one.
>
> RUTH RENDELL, *THE ROTTWEILER*

Computers, then, use *binary* numbers—a *counting base* of two. A counting base is the foundation of any number system, the unit which is multiplied by itself in order to reach higher numbers, as two, four, eight, or ten, one hundred, one thousand . . .

Under our own counting system, based on tens, every time we reach a number which is a power of ten, we start a new column of numbers. The column to the right is for single numbers; the next column to the left is for tens; the column to the left of that is for hundreds . . .

1
10
100
1000

These columns have to do with *positioning* or *place value,* and they make it possible for us to understand, without having to think about it, what a number like 382 means: three hundreds and eight tens and two ones. Three hundred (in the column to the left) eighty (in the column in the middle) two (in the column to the right). We can also read a number like 333 and know without stopping to think about it that it's three hundred and thirty-three, and not three-three-three, a totally different thing.

Binary numbers do the same thing, but here the columns are based powers of two. The best way to understand them is not to think of 0 and 1 in the way that we use them. In this system, 0 and 1 may look like the numerals we use, but they mean something different. Binary numbers don't use a single symbol for the number two, just as in the decimal system, there is no single symbol for the number eleven—we have to use two ones. Binary numbers have to use a one and a zero to mean two.

| | |
|---:|---|
| 0 | (zero) |
| 1 | (one) |
| 10 | (two) |
| 11 | (three) |
| 100 | (four) |
| 101 | (five) |
| 110 | (six) |
| 111 | (seven) |
| 1000 | (eight) |
| 1001 | (nine) |
| 1010 | (ten) |

| | |
|---|---|
| 1011 | (eleven) |
| 1100 | (twelve) |
| 1101 | (thirteen) |
| 1110 | (fourteen) |
| 1111 | (fifteen) |
| 10000 | (sixteen) |

The powers of two are two, four, eight, sixteen . . . and each of those numbers in the chart above starts a new column. There's a rhythm to the count; it's just shorter and bouncier than the rhythm of counting by tens.

For computers, one and zero are everything, the building blocks on which their vast towers of information and communication are built. Everything the computer does, from sending you endless variations on junk mail to solving the most difficult equations, is based on those two symbols, one and zero, and counting by twos.

## IF TWO CAME FIRST

For the Pythagoreans, one was not a number; rather, it was the source of numbers. For them, two was the first number. But there were two things about the number that worried them.

> Doublethink means the power of holding two contradictory beliefs in one's mind simultaneously, and accepting both of them.
>
> GEORGE ORWELL, *NINETEEN EIGHTY-FOUR*

First, they noted that the result of adding two plus two is exactly the same as the result of multiplying two times two. The answer is four in both cases. In fact, any number multiplied by two is the same as that number added to itself. The Pythagoreans thought the mathematics of multiplication ought to be more complicated. (And of course, it is. The simplest thing in math—like multiplying by two—achieves the slightest edge of dizziness when you look at it for a moment too long, somewhat like repeating a word over and over until it becomes nothing more than an inane abstract sound and its meaning whirls away. What's clear can become murky, and what's at the front of your head can recede beyond recall. It takes perseverance as well as clarity of vision to hold on to a number when it wants to tumble away, or to double itself over and over again.)

Early multiplication was a series of duplications. A given number would be doubled as many times as needed until the multiple was arrived at. To multiply 52 by 12, 52 was doubled (52×2=104) and then the answer was doubled again (104×2=208). That number was doubled (208×2=416) and the two answers were added (208+416=624). Thus, 52×12=624. This is a *simple* example of the process. The way we do it is much easier.

The Pythagoreans also worried about two because of the concept of between. Two—which creates between—*has* no between to call its own. It has only a beginning and an end. Something has to be added—one more, for three, they believed, in order to create a between. In that case, the possibility existed that two, like one, might not be a number at all.

## THE DUPLICITY OF TWO

In some ways, the Pythagoreans were right. We can say that Eve was created from Adam, that out of one came two, but only two can make everything else. It takes two.

Eve is the other, the first other, the second. Two. In languages where the word for one and man is the same, another word means both two and woman. Where one means god, two is the transgression that leads away. Eve, woman, two—that step away from God. Two is subliminally a feminine number—and it's an unlucky one. We've inherited that sexist thought without even being aware of it, and we've allowed two to become an unlucky number.

Variations on the bad luck of two include a double cross, which is immoral and dishonest; nobody likes a double-crosser. Nobody trusts a two-faced person. Double trouble is a lot to worry about. Double-talk is meant to deceive, and double-speak is more of the same, deliberately obscure. Double-dealing is back-stabbing. Two-dollar bills come in handy, but no one wants to use them. Duplicity means dishonesty. All are based on the Latin word for two, *duo,* as is *doubt,* which means to disbelieve, to think twice, or perhaps simply the absence of faith.

> If I were two-faced, would I be wearing this one?
>
> ABRAHAM LINCOLN

Happier is a duo, in English—two people, a couple, a pair—and a duet is the music they make. Deuce is more of the same. A diploma was originally two bronze tablets folded together; inside, there was a copy of the law listing the privileges granted to army veterans by the Roman emperors, followed by the name of the individual soldier to whom they were being granted. Eventually, any document listing a privilege, a power, or an authority became known as a diploma. A diplomat carries his credentials in a folder of this sort—one assumes without the bronze—and diplomacy is what he does.

Bi also means two, as bigamy (twice married). Biscotti and zwieback (from the German word for two, *zwei*) are literally baked twice, and that's why they're so crunchy. A bicycle has two wheels; a bison has two horns.

> All human wisdom is summed up in two words—wait and hope.
>
> ALEXANDRE DUMAS, *THE COUNT OF MONTE CRISTO*

The Latin syllables *ambi* follow all of this quite closely. *Ambidextrous* means using both hands equally well; *ambiguous* means having two or more possible meanings; to be *ambivalent* is to have two or more—mixed—feelings.

There is often confusion when *bi-* is linked to time. Does bimonthly mean every other month or twice a month? My dictionary gives both definitions, followed by the note that careful writers avoid using bimonthly to mean twice a month; when that's what they want to say, they use semimonthly instead. Doesn't usually work that way; confusion is widespread. Biweekly and biannual have the same problem; officially, they mean every other week and every other year. Bimonthly periodicals are published every other month. But you will be asked to pay for your subscription at least twice a year, not every other year, and I suspect even more often than that. Your subscription is annual; the plea is probably semiannual.

> Bigamy: the only crime where two rites make a wrong.
>
> BOB HOPE

## ODD AND EVEN AND IN BETWEEN

Mathematically, as one is the first odd number, two is the first even number. Two, four, six, eight—all divisible by two with nothing left over, the hallmark of even numbers, all of which can be divided into two whole and equal parts.

Odd numbers have a different shape to them—they're almost defiant: they won't be divided into two equal piles; they *won't*. You just have to break them apart. Odd numbers are more interesting in that way, because you have to learn a more complicated mathematics to understand them. If five people are sharing one apple, somebody's piece will most likely be largest, unless you divide the apple with a compass. (The old trick for dividing cake into unarguable pieces for two siblings is to let one child make the cut and the other child have first choice of the resulting slices.)

Perhaps because it's the first female number, two has the curves of a woman. Its sensuous line could be a breast, or the smooth bulge of pregnancy, or the mystical rising of a woman's hip as she lies on her side in bed. And two has the mystery of a woman. It is the essence—the very definition—of duality, the yin and yang of numbers. Opposites are pairs of two. Dark and light, top and bottom, good and bad, up and down, true and false, life and death—always two. Heaven and earth, morning and night, the beginning and the end: two. The difference of one's own sex begins with two: man and woman, a twoness, male and female, out of the oneness which is we who are women, or we who are men. Or, if you prefer, we who are neither.

Between the opposites are the shades, the touchings, the gradients. Between is the mystery, the subtlety, the magic. And *between* is a two-derived word, despite the Pythagoreans. To be between is to be in the middle of two things, one on each side, touching both to make two. Between might be a state of uncertainty—neither here nor there, a bit doubtful, the old duo—but between is what is. Twilight—another two-derived word—is the light between day and night, the gloaming, the gradual shading that brings light and darkness together. A branch forks in two, and makes a twig. Tweenies are neither preteen nor teen; they're between. And then there are twins.

The animals entered Noah's ark two by two, pairs of opposites, male and female. Two is the number for opposites, always pulling away from each other, but two is also the number for pairs—those twins, and their parents, and for doubles and duos, always two, coming together, as if opposites, if you stretch them far enough, will touch each other again, like yin and yang, on the other side.

> Round numbers are always false.
>
> SAMUEL JOHNSON
>
> There's luck in odd numbers.
>
> RICHARD LOVELACE

Parallel lines are also pairs, two straight

lines always at an equal distance from each other. Another mathematical pair is the equal sign (=), two short parallel lines, first written by Robert Recorde in 1557, and chosen because, said Recorde, the lines "are twins, and nothing is more alike than a pair of twins."

(Plus and minus were also invented—or chosen—but not by a mathematician. Johann Widman was a businessman in the mid-fifteenth century. His crates of merchandise, according to Denis Guedj in *The Parrot's Theorem,* were all supposed to weigh exactly the same amount. When a crate was lighter than it should have been, he marked it with a dash (–), followed by whatever the short weight was. If it was heavier than it should have been, he put a line through the dash at a right angle (+), followed by the amount it was over the norm. Eventually, he used the same symbols in his accounting books, in order to keep track of the crates, and from there they found their way into addition, subtraction, and algebra. Guedj notes that the minus sign was used before the plus—plus, then, was a minus crossed out.)

## THE POWERS OF TWO

Plus and minus are another of the opposites that combine in the power of two: opposites and pairs, evenness and halves. Equally potent is the mystery and poetry of two. If one comes first, two must always be second. If one is myself, two must always be outside myself, different and other, the beginning of the rest of the world. One is a point, a dot on the page, but two is a connection, a line, from here to there. I am here; you are there; two is between us. Two is we.

> Constantly choosing the lesser of two evils is still choosing evil.
>
> JERRY GARCIA

Our world becomes like the idea of two. We divide it into

twos, into opposites and pairs, you and me, them and us, day and night, east and west, north and south, man and woman, up and down, earth and sky, before and after. But even as we divide it, we bring it together. To separate is to have been one.

You can't think about the idea of two—of twoness—without thinking about the other side, either the opposite or the pair. Two is company, says the proverb. Two is the question and the answer. Two is enough to make more.

# 3

## THREE IS A MAGIC NUMBER

Three is the true start of counting the world. One was the *sine qua non,* me, today, now. Two was the first complication—you, the near, the next. But three is the beyond, the far, everybody else, the rest of the world, and eventually the universe. Two is company, but three is a crowd.

> The step to Three is a step across the threshold of darkness . . . out into the prosaic but clear and bright light of practical life.
>
> **KARL MENNINGER,** *NUMBER WORDS AND NUMBER SYMBOLS*

Children learn early the difference between one and two—the singular and the plural, one and one more than one—but when they begin to go further, they sometimes leap right over three and count one, two, four . . . Three is harder to grasp than two, and that childish leap over the difficulty is really the grasp of a higher concept: that after two comes *many,* and the gap marks the difference.

The Sumerians marked the same difference. Their word *ges* meant man, male, and erect penis as well as the number one. *Min* meant both two and woman. *Es,* three, also worked as a suffix indicating plural, much as *-s* and *-es* do at the end of words in English. Two, then, is a pair, as man and woman; three is where *many* begins. Three marks the plurality, rather than the singular or the dual. In Egyptian hieroglyphics, a quantity is indicated by writing the same symbol three times. In classical Chinese, the idea of a crowd was shown by repeating the ideogram that meant man three times. Three trees meant a forest.

The recognition of *many* is the root of abstract numbers—that one, two, and three are not only linked to whatever is being counted, but rather exist in a sense of their own. Numbers can count anything; they exist apart from what can be touched or seen; they are an idea, an abstract. One becomes oneness.

The roots of the English word *three* are another testimony to the beginnings of the concept of many. The Anglo-Saxon *thri* or *thria* is related to *throp,* which meant a pile or a heap. Words like *throng* are derived from a Germanic word which meant many. Romance languages have the same connections with plurality. The Latin *tres,* three, gives us not only the French *trois,* but also *très*, very, *trop,* too much, and *troupe*—the English troop. In Latin, *tres* is also connected to *trans,* beyond. Another possible link is to the English *through,* in the same sense of beyond. Even *this, that,* and *those* follow the one, two, three pattern. In *The Universal History of Numbers,* Georges Ifrah writes that in Indo-European, the parent language, the word for the number three, *tri,* "was also the word for plurality, multiplicity, crowds, piles, heaps, and for the beyond, for what was beyond reckoning."

Grammar speaks of *many* in a different way. First person is *me;* second person is *you;* third person is everybody else. *I; you; he, she,* or *it,* and, in the plural—*they.* This is the conjugation of numbers in the grammar of our souls. I am one; you are two; he, she, and it—they—are many. *They* is everything that isn't you or me. That's how we order the world—me, you, and them.

## HARMONY

At last, with three, the Pythagoreans had found a number! And what a number they had come to—a number they believed to be a kind of perfection because of the distinction of its parts. For them, one was the beginning; two became the missing middle; and three was thus the end. Three had it all.

> The all and all things are defined by threes; for end and middle and beginning constitute the number of the all, and also the number of the triad.
>
> ARISTOTLE

Since they didn't count one as a number, for the Pythagoreans three was the first male number, given their view of all odd numbers as masculine. It was also their number for harmony, combining the unity of one and the division of two.

For many years, when we first wrote numbers, three was often just three short parallel lines, like one and two, sometimes vertical and sometimes, more often in Asian societies, horizontal. Eventually, people writing quickly with a pen (as opposed to making indentations with a stylus) often joined the lines together, like a backward E, and that evolved into our 3.

Along the way, three was a turning point in the writing of numbers. Many early civilizations wrote the numbers from one to nine as an ongoing series of lines, circles, or dots—something like this:

| I | II | III | IIII | IIIII | IIIIII | IIIIIII | IIIIIIII | IIIIIIIII |
|---|---|---|---|---|---|---|---|---|
| 1 | 2 | 3 | 4 | 5 | 6 | 7 | 8 | 9 |

Once we get beyond four, most of us have to stop and count the lines, a fairly tiresome procedure when we get to six, seven, eight, or nine. The next idea—in Egypt, Crete, and Greece, among other places—was to group the written numbers in two lines, like this:

| I | II | III | IIII | III | III | IIII | IIII | IIIII |
|---|---|---|---|---|---|---|---|---|
| | | | | II | III | III | IIII | IIII |
| 1 | 2 | 3 | 4 | 5 | 6 | 7 | 8 | 9 |

That's a step forward, but it's not an enormous improvement. Other civilizations—the Phoenicians, Assyro-Babylonians, Egyptian-Aramaeans, Lydians—went to a kind of rule of three:

| I | II | III | III | III | III | III | III | III |
|---|---|---|---|---|---|---|---|---|
| | | | I | II | III | III | III | III |
| | | | | | | I | II | III |
| 1 | 2 | 3 | 4 | 5 | 6 | 7 | 8 | 9 |

The rule of three is an example of *grouping*—counting by making groups of things. We do the same thing when we make piles of our loose change—each pile of ten dimes is a dollar; it's easier to count the dollars that way. Tally sticks worked the same way when every tenth notch was deeper than the preceding nine. Counting by grouping makes it possible to go higher than the numbers you know. The number you reach can be visualized, even if it can't be verbalized.

## TRIPARTITE

*Tres,* Latin for three, gave us a host of related words: triple, trio, trident, trinity, triplet, triangle, triad . . . A trice could be the few moments that it takes to count to three. A tripod has three feet; many trivets do too. A tribe was originally the Latin *tribus,* one of the three divisions of the populace (the Romans, the Sabines, and the Albans). A tribute, *tributum,* was the tax paid by each of the divisions. A *tribune* was their choice of someone to protect them from the potential abuse of the patricians. The *tribunal* was the place where the tribune administered justice. More from that same word, *tribune:* tributary, attribute, and distribute. The month of May, a sweet story tells us, was known in the old Saxon as *trimilki,* because in that period of time, cows were milked three times a day.

Another cousin of the Latin *tres* is the Latin word *testis*, or witness—from *tri-stis*, the third person who saw or heard something involving two other people. From *testis*, we have testify, testimony, testament (as a will witnessed by a third person), and even protest—originally, to bear witness against, and contest: to witness against.

## PRIMES AND POINTS

The first three numbers are all prime numbers of a sort. (A prime number is one that can be divided only by one and by itself. Two can be divided by only two and one; three by only three and one.) One would be the first prime, just as you'd expect, except that mathematicians don't consider one to be a prime number at all, perhaps agreeing with Pythagoras about the nature of one. For them, one is neither prime nor composite. (A composite number can be divided into smaller numbers, none of which is one. For example: four can be divided into two and two; both are smaller than four and neither is one.) As neither prime nor composite, one simply *is*. As always, it's in a class of its own. The last mathematician to publicly identify one as a prime—did the others just whisper in classroom corners?—was Henri Lebesque in 1899. For everybody else, then and since, one is considered extinct as a prime, something like a volcano that just doesn't have its stuff anymore.

That leaves two and three tied for honors as the first prime, no matter how confusing that may sound—two is the first even prime, and three the first odd. Some mathematicians consider them to be the Siamese twins of prime numbers because they are the only two prime numbers (and there are infinite numbers of prime numbers) that are not separated by a composite number. They stand there, two and three, joined at the nose with only the customary fractions dancing between their legs.

Three has another mathematical distinction. It's the only whole number that is the sum of all the whole numbers that precede it. That is to say, $1+2=3$.

> If I tell you three times, it's true.
>
> LEWIS CARROLL

Geometrically, three makes a design where before none existed. It takes us beyond the point, which is one, and the line, two. Three is the triangle, the first plane figure. The triangle is determined by three points that are not on the same line—connecting them forms the triangle. The isosceles triangle has two equal sides, and takes its name from the Greek *iso,* same, combined with *skelos,* legs. Other kinds of triangles, with sides of different length, are called scalene—limping. Whether the legs are the same or they limp, the triangle, with its three straight lines and its three rigid angles, is a powerful figure with a fine sense of symmetry and order.

## THE JOURNEY

Three is a knockout as a numeral, with its two curving lines—turned sideways (as nature intended) it's as masculine as the numeral one, making the Pythagoreans early Freudians. You could cup those semicircles in your two hands; they themselves can contain the world—and usually do. Three completes one in terms of masculine symbols—the balls that finish the line.

Surely that's part of the power and magic of three. Three may not be perfect—mathematically, we have to wait for six to find perfection—but it is magic just the same. One is solid, real, and full of substance. Two is sensuous and mysterious, the reverse and the complement of one. Three combines them and stands like a virile colossus over the rest of the numbers.

One is always the *sine qua non,* but by itself, it's just not enough. You can begin with one, and two is enough to continue, but they are not sufficient. You need three to tell the story. One is just the beginning. Three is the journey.

As always, the Romans had a word for it. *Triune* combines the Latin for three and one—*tres* and *uno,* in the religious sense, three gods in one.

The Trinity, the Father, the Son, and the Holy Ghost, is the first of many great threes—not chronologically, but in its impact on the world. The story of Jesus is replete with threes, from the Three Wise Men who came to honor him at his birth to the three crosses at Calvary.

The idea of an all-powerful threesome is not unique to Christianity. For Hindus, the Trimurti consists of Brahma, the Creator; Vishnu, the Preserver; and Shiva, the Destroyer. For Buddhists, there are the Three Jewels: Buddha, the teacher; dharma, the teaching; and sangha, the spiritual community.

The ancients believed the world to be ruled by three gods: Jupiter, who ruled in heaven; Neptune, the god of the sea; and Pluto, god of the underworld. Each of them manifested three: Jupiter holds three-forked lightning; Neptune has a three-tined trident; and Pluto has a three-headed dog.

Have we created threefold gods in a reflection of our own nature—body, mind, and spirit? The power of all of these divine trios surely lies in part in their very threeness, in their sense of three. The children who skip over three in their first countings mark the same concept: one and two become complete when they reach three. Father, mother, and child: three. One, two, family, and family is another *many.* There are so many potent threes—aside from religion, just in the way we view ourselves, and in the stories we tell; there is a sense of roundness, of completion, of magic in the number three. One, two, three, go. Three strikes, you're out. The third time is the charm.

In many ways, we order our world in threes. Like our very selves, with our body, mind, and spirit, our planet is a three: earth, sea, and sky. The kingdoms of nature, we learned as game-playing

> This is the third time; I hope good luck lies in odd numbers. . . . There is divinity in odd numbers, either in nativity, chance, or death.
>
> **WILLIAM SHAKESPEARE,** *THE MERRY WIVES OF WINDSOR*

children, are three: mineral, animal, and vegetable—the only possibility in a round of Twenty Questions.

The spiritual enemies of man are threefold: the world, the flesh, and the Devil. We have heaven, hell, and life on earth. There are three dimensions: length, width, and height. Time, Einstein's fourth dimension, has past, present, and future—three, even as a story has a beginning, a middle, and an end. There are three states of being: liquid, solid, and gas (with only water existing naturally on Earth in all three). There are three cardinal colors: red, yellow, and blue. We eat three meals a day—breakfast, lunch, and dinner.

We divide our very lives in three: birth, life, and death. Or differently, but eventually the same: childhood, adulthood, and old age. Titian painted the three ages of man as first an infant in a cradle, then a shepherd playing a flute, and finally an old man meditating on skulls. Dalí painted infancy, adolescence, and old age. Birth, life, and death, all.

Dante's *Divine Comedy* was written in three parts: the *Inferno, Purgatory,* and *Paradise*. His rhyme scheme throughout was sets of three lines, with the first and third lines rhyming, and the last word of the second line becoming the rhyme for the first and third lines of the next set of three: ABA, BCB, CDC, DED. . . . Each part has thirty-three cantos, not counting the initial introductory canto. Hell has nine circles—three squared—and Paradise has nine spheres before the final Empyrean, where God appears in three circles representing the Father, the Son, and the Holy Spirit.

There are, without a doubt, sets of things for every number. Shakespeare counted the ages of man not at three but at seven; there are four seasons and four phases of the moon. . . . On and on the numbers go, making order of the days of our lives—but no number has the impact and the magic of three. Even Bach counted much of his glorious polyphony as three-part inventions. The Pythagoreans numbered the soul as that which unites the mortal and the immortal, making three, and binding them into a single whole. Three finds the balance between the extremes—the shades and the shadows between light and dark.

## ONLY ONE APPLE

Three stretches from the religiously divine to the ridiculous: Tom Sawyer believed that part of curing warts involved turning around three times. And then it can just as easily go back to the mythic—the story of Paris, whose lust for Helen started the Trojan War. Eris, the Greek goddess of discord, started the trouble. She hadn't been invited to a wedding, and she was miffed. To get even, she threw a golden apple inscribed "To the Fairest" into the celebration, and sat back to watch the fuss.

There was an inevitable lack of agreement about who exactly was fairest—each of the goddesses felt she alone deserved the title. It was finally left to Zeus to make the choice. He was no fool, so he gave the apple to Paris, the son of the king and queen of Troy, and left it up to him to name the fairest. Bribery ensued: Three goddesses, each beautiful and charming, tempted Paris with everything they had. Hera offered wealth and power, Athena wisdom and glory. But Aphrodite, goddess of love, knew best. She said that if he gave the apple to her, he could have the most beautiful woman in the world for his wife. That was the best bait, and Aphrodite won the apple. But as luck would have it, Paris chose Helen of Troy—that she was already married was a mere technicality. Her seduction led to the long Trojan War, and the eventual destruction of the city of Troy and all the myths and legends—like the Trojan horse—that have become part of the story.

> **Transported to a surreal landscape, a young girl kills the first woman she meets and then teams up with three complete strangers to kill again.**
>
> TV LISTING FOR *THE WIZARD OF OZ*, MARIN COUNTY NEWSPAPER

As children, we learn about the powers of three early. Even though there's a one in "Once upon a time," three is the true number for fairy tales. Three is what counts, from the three little kittens who lost their mittens to the three blind mice who lost their tails. There would be no story, just a well-fed wolf, without the three pigs; even the hungry wolf himself huffed and puffed three times. Goldilocks would stand diminished without three—how could anything be just right if it hasn't been too this or too that first? There had to be three chairs, three bowls of porridge, and three beds before she unwisely fell asleep in the one that was just right. There are always three chances. The genie in the bottle, once released, always grants three wishes (none of which can be wishing for more wishes). The king always has three sons, and the youngest always gets three chances in his encounters with strange supernatural beings in the woods, or has three adventures before he finally fulfills his quest and marries his true love—without destroying the city of Troy in the process.

We learn about three as adults, too. In medieval Europe—and still, in some countries, today—there were three main divisions of society: the clergy, the nobility, and the commoners (the last included the burghers or the bourgeoisie, and the peasants). The divisions were called "estates," and formed the legislative bodies or advisory groups to the king, who stood with the numeral one in a class by himself. In prerevolutionary France, the clergy made up the first estate, the nobility formed the second, and the people (everybody else) were the third. The name of the medieval French national assembly, the Estates-General, was based on the three estates; the British Parliament is the analogue. In England, the parallel realms were the Lords Spiritual, the Lords Temporal, and the

Commons, more stately-sounding perhaps than the French, but the same nevertheless.

The estates were essentially male. Women were divided differently, after having been separated simply by their sex. A woman could be virgin, wife, or widow, thus defined by her attachment to a man—with whom a woman is sleeping, with whom she used to sleep, or, poor virgin, with whom she has never slept at all.

## LATER

The fourth estate—the press—came much later than the first three, and it will here too—when we come to four.

But the Third World, as opposed to the third estate, is a much more recent creation. It means, essentially, underdeveloped—or still developing, or in a word of one syllable, poor—countries, and from its coinage came the subsequent First and Second Worlds, though neither expression is used very much. All are a product of the cold war era.

According to Michael Quinion, editor of *World Wide Words,* "Third World" was used for the first time in French—*le Tiers-Monde*—by Alfred Sauvy, a population expert. He was referring to those poor countries, especially in Latin America, Asia, and Africa, that were not part of either the Communist or the capitalist blocs. In an article published in *L'Observateur* on August 14, 1952, he wrote, *"Ce Tiers-Monde, ignoré, exploité, méprisé, comme le Tiers-État"* ("That Third World, ignored, exploited, scorned, like the Third Estate")—a nod toward a well-known pamphlet on the third estate written by the Abbé Sieyés in January of 1789, very much part of what led up to the French Revolution later that year.

Sauvy's use of "Third World" began to be echoed by economists and politicians in Britain and the United States in the early 1960s. "First World" followed in 1967 and meant the developed countries, capitalist and based on high-income market economies, with the United States as the prime example. In 1974, "Second World" followed—the relatively high-income Communist countries, or those in which the government owned the means of production, with the USSR as the best example. The Berlin wall fell in 1989, and by and large, so did the use of "First World" and "Second World." But "Third World" remains—or at least the countries that it comprised are still very much there, only now they're known as "developing nations." Same thing.

## A TROOP OF THREES

There is still a host of threes. There are the three monkeys, See No Evil, Hear No Evil, and Speak No Evil. Even columns come in threes: Doric, Ionic, and Corinthian. For comic relief, there are three Marx Brothers: Chico, Harpo, and Groucho. (Zeppo and Gummo don't count because they weren't really funny.) There were Three Stooges (though a total of six actors played the three parts over the years): Larry, Curly, and Moe. Even Donald Duck has three nephews: Huey, Dewey, and Louie. (In Denmark, they're called Rip, Rap, and Rup; in Finland, they're Hupu, Tupu, and Lupu.)

> **I always have trouble remembering three things: faces, names, and—I can't remember what the third thing is.**
>
> FRED ALLEN

And we've barely begun. There are the three witches in *Macbeth;* the three visible phases of the moon (waxing, full, and waning—the new moon is dark); the three hearts of an octopus (two at each of its gills and a third pumping blood throughout its body); the three eyelids of a camel. (Two provide extra pairs of long eyelashes to help keep sand out of its eyes, and the third moves from side to side, like a car's windshield wiper, to eliminate whatever sand gets through all the lashes. The third lid is very thin, and camels can see through it, so they can keep walking even in a sandstorm.)

We are told that the landmasses on Earth were formed from three original supercontinents: Rodinia, Pannotia, and Pangaea. Babe Ruth wore the number 3 on his uniform. Earth is the third planet from the sun. Captain Kirk and Spock played chess three times in the television series *Star Trek.* Kirk won all three games.

The Greeks and the Romans believed in the three Fates who controlled birth, life, and death—there was one to spin the thread of life; one to measure and weave it; and one to cut the thread, thus ending life. There were three Furies—in Greece, the three Erinyes, whose heads were sometimes shown as wreathed with serpents, while their eyes dripped blood, and who sometimes had the wings of a bat or a bird and the body of a dog. They were Alecto, unceasing; Tisiphone, avenging; and Megaera, jealous and grudging—together they were the spirits of vengeance. On the other hand, there were the three Graces: Aglaia, Euphrosyne, and Thalia, sometimes said to be the daughters of Zeus and Eurynome. They were joy, charm, and beauty, or, if you prefer, splendor, festivity, and rejoicing. The Muses were three times three (which takes us to nine, another magic number, which we shall soon reach). And there are the three Christian graces: faith, hope, and charity.

The triple sacrifices are milk, wine, and honey. The augurs were consulted three times, and Pliny spat three times to be sure that a dose of medicine would be effective. The cock crows thrice: it's dawn.

Luck, both good and bad, comes in threes. (Bad: three cigarettes on one match, because during World War I the long-lasting light gave the enemy time to sight, aim, and fire.) We say that accidents—especially plane crashes—come in threes. We offer three cheers for the winner—hip hip hooray!

But more than anything else, three is most intimate, magical, and powerful because three is the number for a child. Mother, father, and child make three, and so three is the number for the future.

# 4

## FOUR IS A SOLID NUMBER

Four goes gladly through the gate that three opened: we are now solidly into *many.*

*Many* is an idea that comes up again and again when we think of numbers, from the most primitive countings of one-two-three-many to the complicated words we use today when we mean *many:* millions of this, thousands of that, hundreds of the other—all ways of expressing the enormous, ineffable idea of *many.* And there are so many nonnumerical ways to say *many:* Think of gobs, heaps, loads, lots, oodles, tons, scads, a slew, a passel, a wad, a bushel, and a peck.

> Four ducks on a pond,
> A grass-bank beyond,
> A blue sky of spring,
> White clouds on the wing:
> What a little thing
> To remember for years—
> To remember with tears!
>
> WILLIAM ALLINGHAM, "A MEMORY"

And then there are the collective phrases—the words that describe *many* without exactly saying so, and without referring to number. Now that we know how to count, collective phrases have turned into poetry—without rhyme, without rhythm, but poetry because they are filled with the double wonder of sound and of image.

The first English collection of these phrases was in *The Book of St. Albans,* published in 1486, probably based on earlier works written in French. It had three parts, on hawking, hunting, and heraldry; when it was republished (because it was so popular), a new section—*Treatyse on Fysshynge with an Angle* (rod fishing)—was added.

Many collective phrases describe animal behavior. A parliament of rooks is about the way the birds chatter in their nests in tall trees; an exaltation of larks is a poetic description of the climb of skylarks, soaring into the sky as they sing; a murmuration of starlings tells about the twittering of those birds as they roost in the early evening; and an unkindness of ravens refers to the old belief that ravens push their young out of the nest to survive in any way they can. Those are all birds; there's also a pride of lions, a parade of elephants, and a tower of giraffes.

Many other collective phrases speak of bunches of things—more than a brace, but a gathering without number: a swarm of bees, a herd of sheep, a flock of birds, a pack of dogs, a string of horses.

Some collective phrases sound strange to our ears: a descent of woodpeckers, a plump of waterfowl, a tiding of magpies, a pitying of turtledoves. Some are funny: a shrewdness of apes, a prickle of hedgehogs, a bloat of hippopotami, a passel of possums. Some just sound right—either to the ear or the mind: a mischief of mice, a pit of snakes, a scurry of squirrels, an ambush of tigers. The easy favorites are the ones that sound most like a poem: a glint of goldfish, a down of hares, an aerie of eagles, a charm of hummingbirds.

Some words are used over and over in collective phrases. A pod can be a gathering of dolphins, penguins, or whales. A school can be just plain fish or, individually, salmon or porpoises. A herd can be a variety of animals, and a swarm can be more than bees—eels, flies, even rats. Hordes can be small animals or enemy soldiers. Packs can be wolves or dogs, and a deck of cards—the last its own collective phrase. Armies are ants and other bugs; colonies are badgers or beavers; a bevy works for beauties as well as for doves; a bouquet was pheasants before it was flowers. But there is only one collective phrase for each of these: a scold of jays; a peep of chickens; a storytelling of ravens (does a storytelling of ravens differ from an unkindness only in the mind of the beholder?); a clowder of cats; a gaze of raccoons; a shiver of shark.

Some animals have more than one phrase to describe their behavior: a gaggle of geese is on the ground; a skein of geese is in the air; and a wedge of geese is flying in a V-formation. In the same way, ducks waddling on the banks of the river are a team of ducks; swimming in the stream, they're a paddling of ducks.

All the collective phrases sound made up in their perfection, and of course, they were—some of them longer ago than others—but they all speak of groups of things without specific numbers. Most have a magical, unexpected sound to them—which is exactly the way in which they sound contrived, as if they had been written by a poet—even though when we stop and think about them, we can usually see fairly clearly why they are as they are. There are hundreds of them, some so familiar that we take them for granted—a bed of clams or oysters, a flight of birds, a host of angels, a flood of tears, a bouquet of flowers, a bed of roses, a flight of stairs—and many more that are rarely used today—a sloth of bears, a brood of chickens, a murder of crows, a sounder of swine, a skulk of foxes, a party of jays, a kindle of kittens, a plague of locusts, a clutter of spiders, a drift of swans. . . .

Even people have their own collective phrases: a troupe of dancers, a team of athletes, a slate of candidates, a panel of experts, a band of men, a gang of thieves, a crowd of people, a crew of sailors, a troop of soldiers, a coven of witches, a board of directors, a cast of actors. And things have more: a quiver of arrows, a batch of bread, a bunch of bananas, a fleet of ships, a clutch of eggs, a chain of islands, a range of mountains, a stand of trees, a fall of snow. . . .

They're all about numbers, and they're all about *many.*

## AND MANY MORE

If three and four stand as the beginning of true counting, *many* is, in a way, its end—not infinity, but the closer and more diffuse concept of a multitude. At the beginning, counting was more concrete—there was something here to be counted, something specific and real. *Many* is different; it floats off into numberless clouds, an array that is real but without exact dimension.

Some Oceanic languages have separate words for one, two, three, and four of whatever is being counted—and then a word that means *many.* The Brazilian Botocudos had only two number words, for *one* and *pair.* They went a bit higher by saying *one-and-pair* for three and *pair-and-pair* for four, but the idea of anything after that was remote. (To say *many,* they pointed to the hair on their heads.) Many other languages had words for only one and two, as the Australian *urapon* and *ukasar,* where three becomes *one-two,* and four is *two-two.* The numbers stop at that point, and the next word is *many.* It's not hard to imagine counting that way: I can have one car, or two, or maybe even three or four, but beyond that, I just have a lot of cars—*many* (and somebody to take care of them). We say things like "I have a million things to do today," but we don't mean that literally; we mean *many.* Where those *one, two, one-two, two-two* languages are spoken, someone visualizes things—goats, perhaps, or fish—in the same way. More than four is simply a lot. *Many.*

> There comes a time in every man's life and I've had many of them.
>
> CASEY STENGEL

More complicated on the path that counting took as it grew, there were people who counted by using specific images rather than numbers—like one-to-one counting, about as concrete as you can get. *Bird-wings,* of which every bird has two, works for that number; *wings* and *eyes* were used in many cultures to mean two. The Caribs, on islands in the sea that was named after them, used a word for two that meant break or split—to show that two was created out of one. Three was sometimes *clover-leaves,* and four a word like *animal-legs,* because most animals (other than people) have four legs. Five, if things got that high, was, most obviously, *fingers-of-one-hand.* If you had five jars, you had *fingers-of-one-hand* jars. Higher numbers were more difficult, and not just in terms of memory. Spiders have eight legs, but what could be used for nine? (Eleven was sometimes *one-at-the-foot,* the number that came after all the fingers were used and the toes were being added.)

> Our perfect companions never have fewer than four feet.
>
> COLETTE

For others, counting words were more complicated. For them, there were separate sets of words, each to be used for the kind of thing being counted. Flat objects had one set of number words; round objects had another set. Long things like trees had their own set, but other long things like canoes had another. There were sets of words for measuring, and a separate set to use for people. A final set covered anything that wasn't included in one of the other sets.

The Yurok Indians, in northwestern California, had fifteen separate ways of counting, depending on what was being counted. People had one set of number words, houses another. Boats and long skinny things (like snakes and ropes) each had their own set, as did trees and, separately, plants other than trees. Tool, rocks, round objects (like coins)—each had its own set. That makes for a lot of number words, a great deal to remember. Counting that way can't go very high.

Still, Malcolm Margolin wrote in *The Beauty of Disconnection* that this coding gives us a glimpse of the way in which a language can capture a kind of truth. "It is quite clear that the sevenness of seven human beings is qualitatively different from the sevenness of seven trees, both of which are qualitatively different from the sevenness of seven birds, and so on. How precise this language is, to pay tribute to distinctness and to force those who speak that language to acknowledge the incomparability of different things. Our own system of counting, one-size-fits-all, has produced enormous benefits. . . . Yet I can't help but feel that among all the gains, we have lost something . . . a manner of thinking that allows us to see the world as something composed of unique, distinct, incomparable entities and small systems of great variety, great individuality, with rugged strengths and inviolable character."

There have been others, like the Yuroks, who used the same idea—separate counting words for sets of things being counted—but went about it differently. They had one word for *two* long things, like trees, but when they counted *three* long things, they used a completely different word. In the flat-thing category, like leaves, there was one word for two leaves, and an altogether new word for three leaves. Then two birds had one word, and three birds had another. Again, too much to remember; counting that way can only go so far before memory dissolves. *Many* fills the gap. One bird, two birds, three birds, four birds, *many* birds. *Many* is an abstract idea. Eventually, the numbers gave up their concreteness and joined it.

It seems so simple to us now. Three is three, whether it's long things or short. It doesn't matter if a tree looks different from a bird; what matters is how many you have of each. The idea of a number means letting it float away in your mind and knowing that it will still be there when you come back with things to count, and knowing at the same time that a number can exist nowhere but in your mind. Whatever you're counting is real—it may very well be touchable, unless you're counting something like the dreams you've had since Monday. But numbers are not real; they're an idea. Abstract numbers are a little like bubbles. If you look at them too long, they're gone. You have to trust that they're still there, even—especially—that they exist at all. Four, after all, means four, and it always will. And it can count anything, babies or bananas. Round things don't matter, and neither do flat or long things. Four is four and that's all there is to it. What's amazing is how sophisticated that simple idea is.

## WRITING FOUR

It's possible to count to four by glancing at a group of things. For most people, it's difficult—or impossible—to go higher. **** is a row of four asterisks—you can tell without thinking about it, without counting. But ******* is just a bunch—too many to tell how many without stopping to enumerate.

The first number forms for one, two, and three are ancient and simple—straight lines that are themselves a count, one line for one, two for two, three for three. Four is different. Four lines are just that much more complicated and tedious to write and to comprehend.

For a while, four endured with its row of straight lines, piling one on top of another or standing them in a row like so many dominoes waiting to be toppled. The first difference came in India, about two thousand years ago. Four began to be written like our plus sign ( + ) a horizontal line crossed at a right angle to indicate four. It's graphically appropriate, with its four arms, or tips. That form evolved through a series of loops at its various ends and eventually it became more tiresome to write than the four straight lines had been.

Arab mathematicians wanted a faster way to write the numeral for four than the symbol which had reached them from India, so they eliminated a lot of the fancy work. They scrapped most of the cursive curls and went back to the plus sign. They wrote it without lifting the pen from the paper, giving it a curve between the top of the vertical and the beginning of the middle horizontal on the left hand side, and adding a little flourish at the end. The next step came after Arabic numerals had crossed the sea into Europe: the final curve was dropped and what was left was the connected cross. Eventually, the curved connection became straight, and our triangular four (4) was born. It had evolved, but its genes were intact. It was now a three-sided triangle, with a dip at the bottom, like an onstage bow, to make four.

## FOUR-SQUARE

Our other written four replaces the triangular figure with two sides of a square, the left side and the bottom, but keeps the plus sign intact on the bottom. That rectangular four matches its parent words in Latin, *quadra,* "square" or "cross," and *quattuor,* "four," and from both we have two related words in English: *square* and *quarter.*

> **I recommend you to question all your beliefs, except that two and two make four.**
>
> **VOLTAIRE**

We think of squares, with their four equal sides, as being strong and self-contained: perfect. A square deal or a square meal is simple, fair, normal, and satisfying.

We use these comforting squares as a symbol for home—children draw houses as four-walled squares, seen from the front with a door in the middle, windows on each side, and a chimney on top.

Squares are also the symbol for the very heart of a city. Old cities were divided into quarters (*quattuor* again)—four major areas, with the city square at the center, an area of grass or a park, surrounded by buildings, one of them perhaps a cathedral, or the city hall.

A quarter is one-fourth. There are four quarters in a dollar, and four quarts in a gallon. A quadrant is a fourth of a circle. A quarter of an hour is fifteen minutes; four times fifteen is sixty, and the small hand moves.

Quarantine once lasted for forty days—four times ten. The word *quarantine* comes from the Italian *quaranta giorni,* "forty days," but the first record of a quarantine is from the city of Dubrovnik, where according to documents dated 1377 in the city's archives, newcomers had to spend a period of isolation on a nearby island, to see if symptoms of the dreaded plague would emerge. In the fourteenth century, plague—the Black Death—killed between a quarter and a half of all of the residents of Europe, so quarantine was a good idea. There have been many other quarantines since, but they all have originated with the number four and most have lasted for forty days.

> Two and two continue to make four, in spite of the whine of the amateur for three, or the cry of the critic for five.
>
> **JAMES MCNEILL WHISTLER,** *THE GENTLE ART OF MAKING ENEMIES*

More squares and quarters: Quarries once produced square blocks of stone. A squad was originally a square formation of soldiers. The Italian *squadra,* "battle square," was enlarged in the sixteenth century to become *squadrone;* in French the word became *escadron;* and in English, *squadron.* Headquarters is where the four quarters come together; it's the center; the head, one imagines, where the brains are supposed to be.

You can see the word *four* more directly in the English word *fortnight,* which has come to mean fourteen days or two weeks. It comes from the Middle English *fourtenight,* "fourteen nights," which in turn is descended from the Old English *feowertene*—"fourteen"—coupled with *niht,* "night." Fourteen and forty have four on clear display, and the French *quatre* does the same for the Latin *quattuor* (as do the Italian *quattro* and the Spanish *cuatro*). The Indo-European prototype is *kwetwor,* a similar sound. The ancient Greek for four was *tetra, téttares, téssares,* or *tétores,* from which we have another host of four words: a tetrahedron is a solid with four triangular faces; one example of a tetrahedron is a pyramid. (The Pyramids of Egypt have square bases and four triangular faces sharing a common vertex.) A tetrarch was the ruler of one-fourth of a province; a tetrameter is a line of verse with four metrical feet. Would you like to know that a firkin is an old diminutive for four, and that there are four firkins in a barrel? Or that the old British farthing, a quarter-penny, descended from the old Anglo-Saxon word *feorthing,* for a small fourth?

## COUNTING THE GAPS

At various times and in various places, many people learned to count by fours. When we look at a hand now, we see five fingers. But for a long time, we didn't count the thumb as a finger equal to the others, and so the count went from one to four on one hand and then from five to eight on the other. We still measure the height of horses by "hands." A hand in that sense is the distance across one's knuckles, and doesn't include the thumb.

> Whatever a man prays for, he prays for a miracle. Every prayer reduces itself to this: Great God, grant that twice two be not four.
>
> IVAN TURGENEV, "PRAYER"

In a related way, there were places where people counted by four differently. They used the gaps between their fingers as their measures, rather than the fingers themselves. We look at our hand and see five fingers; they saw four gaps. Some North American Indians counted in that way—four gaps on each hand; eight on both. A visual way of making the count go higher was to place sticks in the gaps, holding the fingers tightly together to keep the sticks in place. The first stick meant nine (one after eight, having used both hands); the second meant ten. A stick in every gap on one hand meant twelve; sixteen needed one stick in every gap on both hands. The numbers (four-based) went only to sixty-four and then stopped. More sticks than that just wouldn't fit. Any number after sixty-four was the elusive, indefinable, and ever marvelous *many.*

## FOUR POINTS

Geometrically, four makes a solid. One is a point, just a dot, on a blank page. Two is a line, the connection between two points. Three is a surface, a design, a circle or a triangle that follows the dots from place to place. But four offers more, as the dot suspended over a triangle turns it into a tetrahedron when it's connected to the triangle's three corners.

> I know that two and two make four—& should be glad to prove it too if I could—though I must say if by any sort of process I could convert 2 & 2 into five it would give me much greater pleasure.
>
> GEORGE GORDON, LORD BYRON, IN A LETTER TO ANNABELLA MILBANKE

Dividing a circle into four—like a pie meant to serve four people—makes four right angles where the lines cross in the middle. Circles, no matter how small or large they are, have only one form—unlike quadrilaterals, which have four sides but can be rectangles, squares, trapezoids, rhombuses, or parallelograms. Quadrilaterals have four sides and four angles, and the sum of their angles is 360 degrees.

The word *four* is the only number in the English language which has the same number of letters as the number itself. . . . One and two have three letters each; three has five; but four has four. Not of great significance, I grant you; just nice.

Without a doubt (aside from anything else, they didn't speak English), that's not why the Pythagoreans believed four to stand for the idea of justice. Rather, it was because the number four is the sum and product of equal parts—two and two—and because squares, with their four even and equal sides, are perfect figures. Is justice perfect? Or equal? No, on both counts. But it should be.

Numbers have double (and often hidden) meanings in different cultures. Thirteen is unlucky in the United States; in Chinese, the number 250 is slang for stupid, so it's avoided, but four is China's thirteen—the unluckiest number of all. *Si,* "four," also means "death" in Chinese. In Japan too, four is not a good number. Garages have no parking spaces with a four (as American buildings have no thirteenth floor); there is no seat number four on a Japanese plane, nor is there a room number ending in 04 in a Japanese hotel. The superstition goes back to a coincidence of sound that began with the adoption of the Chinese number system and the resultant rules for reading and writing. The Sino-Japanese word for four is *shi,* and, like *si* in Chinese, that sounds like the Japanese word for "death." Instead, the Pure Japanese word *yo* is used as the name of the numeral four. Four men are *yo.nin,* because *shi.nin* can also mean "death" or "corpse." To say "seven," the Japanese avoid the Sino-Japanese *shichi,* because *shichi* sounds like *shitou,* which means "death" or "loss"; they say *nana,* the Pure Japanese word instead. Americans consider forty to be a dangerous age for a man—the year of the midlife crisis, worse than the seven-year itch—but forty-two is the danger age for Japanese men because the word for forty-two is *shi.ni,* again sounding like "death"—and because the word *shin.i* means "occurrence of death," and the sound is much too close to be safe. The word becomes *yon.jû.ni,* but it's still a dangerous age.

For many, four is exactly the opposite of what it is in Japan. The Latin *vena amoris,* the "vein of love," was the vein that supposedly ran straight from the ring finger on the left hand to the heart, and so the fourth finger is the place for a wedding ring, a circle of love. As Swinburne wrote in 1680, "The finger on which this ring is to be worn is the fourth finger of the left hand, next unto the little finger; because by the received opinion of the learned . . . in ripping up and anatomising men's bodies, there is a vein of blood, called *vena amoris,* which passeth from that finger to the heart."

In addition to believing that four stood for justice and equality, the Pythagoreans also added the concept of order to the symbolism of the number four, because four is necessary to determine where we are in the universe, which they divided into four cardinal points: east, west, north, and south.

Four measures time as well as direction. There are the four seasons: winter, spring, summer, and fall, revolving as the year goes by. The months see the four phases of the moon: new, waxing (first quarter), full, and waning (last quarter). And four counts out our days: morning, noon, afternoon, and evening.

Four has always been a number significant to the earth—many early cultures believed in the four winds, and sometimes in a four-eyed, four-eared god who supervised creation. For the ancient Egyptians, the cosmos was sometimes shown with a heavenly roof resting on four supports—pillars or mountains or women—at its corners. In Hebrew (one of the early languages in which letters doubled as numbers), the word for Adam was also a symbol for the number four.

> **When speculation has done its worst, two and two still make four.**
>
> SAMUEL JOHNSON, *THE IDLER*

Once, we believed that our own bodies were based on four humors or energies, each related to body fluids that determined our health and dispositions. The sanguine humor related to blood and cheerfulness; the choleric humor to bile and anger; the phlegmatic humor to mucus and apathy; and the melancholic humor to black bile and depression. Outside our bodies, we believed that the world was made of four elements—earth, water, fire, and air.

The Pythagoreans, ever defining in terms of numbers, believed that knowledge could be divided into four studies, the quadrivium, or four paths—a crossroads of knowledge. The first study of the quadrivium was arithmetic; the second was the study of numbers or music—music as the ancient name for the mathematical study of ratios. The third study was geometry, the study of the three dimensions: length, width, and height. Astronomy was the fourth study. All four are really about number: arithmetic is pure number, the study of multitude and quantities. Music is number in time; geometry, number in space; and astronomy, number in space and time. The quadrivium totaled one part of knowledge; the other, the trivium, from the Latin for three roads—*tri-via* (yes, "trivia")—presented the three basic disciplines: grammar, rhetoric, and dialectics. Grammar was considered the mechanics of language; dialectics, the mechanics of thought; and rhetoric, the use of language to teach or persuade. (A trivial bit of lore: Rome's Trevi Fountain, with its three streams of water, used to be called Fontana Trivia.) Until the fifteenth century, the undergraduate degree was equivalent to the trivium; the graduate degree, the quadrivium. Add the quadrivium and the trivium together—four plus three—and you get the seven liberal, and lively, arts.

## DANCING BY FOURS

The quadrille took its name from an eighteenth-century French card game for four players; it became a dance for four couples, one on each side of the set—like American square-dance forms. The dance was livelier than the stately minuet, but it had nothing on the waltz.

The basic tempo of music is four-four (four beats to each measure of music). Four-four time has the solidity and evenness of the number four itself. Marches are in four-four time, and so are most slow movements—adagios and largos. But waltzes—ah, waltzes—are in three-quarter time (there are three beats to each measure, and the main accent is heard every three beats instead of every four; the quarter of three-quarter time refers to the basic time signature), and they have the lilt and sweep to prove it. They're lighter and as a rule faster—your feet want to tap (not to mention dance) when you hear one.

The waltz grew out of the *ländler,* a German country dance in three-quarter time. The word *waltz* is from an old German word, *walzen,* which meant "to roll," "turn," or "glide." The *ländler* involved a certain amount of countryish hopping and jumping, but the waltz polished all that with a gliding grace that had enormous appeal for the young men and women who first danced it.

**If there exists a form of music that is a direct expression of sensuality, it is the Viennese Waltz. . . . The contemporaries of the first waltzes were highly shocked at the eroticism of this dance in which a lady clung to her partner, closed her eyes as in a happy dream, and glided off as if the world had disappeared. The new waltz melodies overflowed with longing, desire and tenderness.**

MAX GRAF,
AUSTRIAN MUSIC SCHOLAR

It also meant that the man held the woman in a light embrace as they danced—scandal! The aristocratic dances that preceded it—the minuets, polonaises, and quadrilles—kept the partners at a decorous distance from each other. But waltzers had to hold on to each other as they swooped around the dance floor. The waltz was banned in parts of southern Germany and Switzerland. Religious leaders everywhere considered it to be vulgar and sinful.

When the waltz reached England, the *Times* of London considered it necessary to write an editorial about its lasciviousness. "It is quite sufficient to cast ones eyes on the voluptuous intertwining of the limbs and close compressure on the bodies in their dance, to see that it is indeed far removed from the modest reserve which has hitherto been considered distinctive of English females. So long as this obscene display was confined to prostitutes and adulteresses, we did not think it deserving of notice; but now that it is attempted to be forced on the respectable classes of society by the civil examples of their superiors, we feel it a duty to warn every parent against exposing his daughter to so fatal a contagion." Onward to the tango!

Four, even aside from the risky waltz, does have its ways. Four is a four-letter word, and four-letter words can raise a few eyebrows here and there. In some technological circles, *four,* all by itself, is a synonym for the prime four-letter word. That usage derives—as numbers so often do—from counting on one's fingers. In the binary system (remember, computers count with binary numbers) several numbers can be shown by raising the middle finger of one hand—the classic gesture. Things can get worse: there are binary numbers that use the middle finger of both hands at the same time. On the other hand (not that there are three) there aren't many people who count binary numbers on their fingers.

## BACK AT THE ESTATE

The three estates—the church, the nobility, and the commoners—were to be joined by a fourth. In 1841, Thomas Carlyle wrote, "Edmund Burke [the Irish-born English writer and political thinker] said that there were three Estates in Parliament, but in the Reporters' Gallery yonder, there sat a fourth Estate more important than them all." The fourth estate has been the press—and now the media—ever since.

For a while, though, the fourth estate referred not to reporters and editors but to something perhaps even more dangerous. In 1752, Henry Fielding wrote, "None of our political writers . . . take notice of any more than three estates, namely Kings, Lords, and Commons . . . passing by in silence that very large and powerful body which form the fourth estate in this community . . . The Mob." Does that give us five estates? Church, nobles, commoners, press, and the mob? No; the fourth estate remains: newspapers, magazines, television, radio, even the Internet. The news. The mob can settle for third.

Four and its offspring, forty, are powerful in other ways as well. Four has religious reverberations—much less than three, but still a surprising amount. For Christians, the New Testament opens with the four gospels of Matthew, Mark, Luke, and John—the Four Evangelists. At the end of the New Testament are the Four Horsemen of the Apocalypse: Conquest, on a white horse; Slaughter, on a red horse; Famine, on a black horse; and Death, on a pale horse—pale horse, pale rider.

In Buddhism, there are the Four Noble Truths: that all existence is suffering; that this suffering has a cause; that the suffering can be suppressed; and that there is a way to accomplish this, through the Noble Eightfold Path, which is twice four (and which we shall read more about when we reach eight).

Forty is in its way another kind of stand-in for *many*—a vast and large number that in time became more specific. Lent is the forty-day period before Easter, a time of reflection and repentance that ends with Holy Thursday, Good Friday, and Easter Sunday.

In the Old Testament, the children of Israel wandered in the wilderness for forty years; Moses stayed on Mount Sinai for forty days. When the deluge came, it rained for forty days and forty nights.

Less seriously, in England the weather on Saint Swithin's Day forecasts forty days of rainy or dry weather—a bit like American Groundhog Day, but holier, the Punxsutawney hole notwithstanding. Saint Swithin wished to be buried in a churchyard, where "the sweet rain of heaven" would fall on his grave. And so he was—at least until he was canonized, when the monks decided to honor him by moving his body into the church. They set a summer day—July 15—for the move, but when that day arrived, it rained. And rained. And rained. It took forty days of rain for the monks to realize that their saint was not happy about being moved inside, and they abandoned the project. July 15 remains Saint Swithin's Day, and the English hope for dry skies on that day.

> How, if on Swithin's feast the welkin lours,
> And ev'ry penthouse streams with hasty show'rs,
> Twice twenty days shall clouds their fleeces drain
> And wash the pavement with incessant rain.
>
> JOHN GAY, "TRIVIA"

English common law set the original privilege of sanctuary as a period of forty days. An English widow was once allowed to remain in her late husband's home for forty days—a historically sexist sanctuary. Strangers could remain in one place for forty days before they'd be required to tithe there. Members of Parliament were protected from arrest for forty days before Parliament stopped meeting, or for forty days before it met, providing political sanctuary for august corrupters.

There are many more fours. The four rules of arithmetic are addition, subtraction, multiplication, and division; there are four suits in a deck of cards; most forks have four tines; the Four Corners in the United States is the only place where you can stand in four separate states—Colorado, Utah, New Mexico, and Arizona—at the same time; people in animated cartoons have four fingers; our hearts have four chambers; and human beings have four blood groups.

> The reason why so many people never get anywhere in life is because when opportunity knocks, they are out in the backyard looking for four-leaf clovers.
>
> WALTER CHRYSLER

There are other forties as well. Ali Baba, for one, met with forty thieves. And finally—but not really, because there's always more—alchemists believed that it would take forty days for their magic changes to take place, for the philosopher's stone to appear and change base metals into gold, and for the elixir of life, the cure for all ills, including old age and death, to emerge. Alas, it turned out that forty days were not enough. There's still no cure for even the common cold.

# 5

## FIVE IS FOR FINGERS

Five, says the dictionary tersely, is one more than four. Well, yes. By the same token, it's one less than six. But that doesn't begin to tell the story. Five is a world ahead of four in what it can do. Five is power; five is fingers curled into a fist; five is a punch in the nose.

The Sanskrit word for five is *panca,* from which, in a straight line, we have our word *punch,* which is not only what a boxing glove does; it's also a drink originally made with five ingredients: arrack, lemon juice, sugar, seasonings, and water.

Dictionaries also tell us that the word *digit* can be two things: any number from zero to nine, and a finger. (*Digit,* you remember, is from the Latin word *digiti,* "fingers.") Using the same word for numbers and fingers tells us what is perfectly obvious: we used our fingers to count for a very long time. Who wouldn't? Counting with your fingers is a long way from moving your lips while you read; counting with your fingers is easy, fast, and natural. It's only logical that it should have come first, before numbers. We've just added more complications to the machinery, is all—because we needed permanent records, and ways of communicating with people who couldn't see us.

> A child of five would understand this. Send someone to fetch a child of five.
>
> GROUCHO MARX

We still use our hands, in a way, to count. When we want to

tell someone a number without saying anything aloud, the easiest thing to do is just to hold up a hand, with the fingers slightly spread out, make a fist, hold up the hand again, and keep doing that, counting by fives until the correct number is reached. "How many?" the hostess in the restaurant asks when we come in the door. Two, we say—and hold up two fingers just to be sure.

Children shouldn't be stopped from counting with their fingers, even if it does seem as if they're cheating, in a way, or slowing themselves down. They're not; they're internalizing numbers, and they're learning how organic numbers are, how related to our bodies. They're also getting a sense of numerical progression and incorporating the idea of number placement. We write numbers from left to right. As adults, we take that for granted, but as children, somewhere along the way, we have to learn, and it helps—perhaps subliminally—to have a body feel for right and left, and for where the numbers are in relation to each other. It helps to have that at the tips of your fingers first.

> What this country needs is a good five-cent cigar.
>
> **THOMAS R. MARSHALL,** REMARK TO J. CROCKETT, CHIEF CLERK, U.S. SENATE

> There are plenty of good five-cent cigars in the country. The trouble is they cost a quarter. What this country needs is a good five-cent nickel.
>
> **FRANKLIN P. ADAMS**

The fingers at the ends of our hands at the ends of our arms tell us two things. They answer the basic question, *How many?* And they come in order, so we know what's first and what's fifth. They give us the succession of numbers. They are our first counting machines, and they remain the simplest and most natural. All that, and they're portable, too.

They also tie into the rule of four—that we have the natural ability to see and count four things at a glance. We have four fingers in a group on our hands, and then our marvelous thumbs off to the side, and so we grasp not only the world, but one more number: five.

## COUNTING BY HANDS

Finger counting is much more complicated than it seems to us. We think of children using their fingers to add four and three, or two and seven, but true finger counters could do much more than that. They saw the parts as well as the whole. For them, each finger was composed of three sections connected by two joints (except for the thumb, with only two sections and one joint), and in a pinch, they moved beyond the fingers themselves to the knuckles. The grand total of those parts came to more than twenty-five on each hand, plus the original five. Then the fingers could be bent in various ways—to reach the top of the palm or the bottom or dangle in air. The thumb could not only be bent as it stood, but could also stretch across the palm in various positions.

It also mattered which finger you used to count the others. Turks counted by using one thumb to count, in this order, the tips of the fingers, beginning with the little finger, for a count of four. To show five, the thumb was raised. To proceed, the thumb was bent and the count went backward, from the index finger to the pinkie, and last of all the thumb again. At ten, all the fingers of one hand were stretched out to indicate that they had been counted. Older methods raised one finger of the other hand for each ten that had been counted; this hand with the fingers splayed out meant a total of fifty.

A tribe of Native Americans (the Déné-Dinje) used words to finger-count numbers. *The end is bent* meant one, and to show it, the little finger was bent. *It is bent once more* was two, and this time the ring finger was also bent. *The middle is bent* added the middle finger for three, and four was a poetic *only one remains.* Equally poetic: *my hand is finished* meant five. Two trees, then, would be *It is bent once more*—the words standing for the number. Other peoples in other places used *my hand dies* instead of *my hand is finished* for five (and my *hands are dead* for ten, *my hands are dead and one foot is dead* for fifteen, and *a man dies* for twenty).

The Greeks finger-counted, and so did the Romans. (The Romans said *Novi digitos tuos,* "I know your fingers," when they wanted to compliment someone's reckoning skills.) In fact, traces of finger counting have been found almost everywhere in the world—Polynesia, Oceania, Africa, Europe, Egypt under the pharaohs, the Middle East, China, India, native North America. . . . Wherever they were, fingers were counted upon. Only the methods varied. Sometimes the counting began with the fingers closed into a partial fist; other times they stood open. Sometimes the counter began with the little finger; other times with the index finger; sometimes with the thumb.

In the Middle Ages, there were scholars who could count from one to ninety-nine on their left hands alone. They had to remember all ninety-nine positions, but when they did their system worked. Then they counted from one hundred to ten thousand—ten thousand!—by making the same motions with their right hands. And they could go even higher by using both hands. The Venerable Bede, eighth-century scholar, historian, and theologian, said he could count to a million on his fingers—but surely, he didn't have to do that very often. Keeping track of all those fingers and thumbs, joints and knuckles, was not a simple process, and indeed, the ability to finger-count was once the mark of an educated man. Manuals of arithmetic included careful instructions on how to count with the fingers.

The Chinese counted to billions on their fingers. They counted three sides to each finger, knuckle, and joint—the left, middle, and right. Each finger then counted to nine; the little finger of the right hand counted units; the ring finger, tens; the middle finger, hundreds; the index finger, thousands; the thumb, tens of thousands. And that still left the fingers of the left hand. The thumb on that hand represented hundreds of thousands; the index finger, millions; and so on to hundreds of millions—billions. Incredible.

Some Hindus count months as having fifteen days—half of a full lunar month, either the waxing or the waning phase. The days of each phase can be counted on one hand. When beads aren't available, some Muslims use their fingers to count the ninety-nine incomparable attributes of Allah, or for counting the *subha,* the ninety-nine repetitions that follow the obligatory prayer.

Finger numbers eventually began to have meanings beyond pure numeration. Saint Jerome wrote that the finger-counting sign for thirty meant marriage. The tip of the index finger was held against the tip of the thumb—thus, he wrote, "a tender kiss represents the husband and the wife." Sixty, with the index finger bent over the thumb on the left hand, so that its second joint touched the thumb's tip, represented a widow "in sadness and tribulation." One hundred ("pay close attention, gentle reader," he wrote), used the same fingers on the right hand, with the tip of the index finger touching the bottom section of the thumb and the space between them closed, "shows the crown of virginity."

> The main facts in human life are five: birth, food, sleep, love, and death.
>
> E. M. FORSTER, "ASPECTS OF THE NOVEL"

Less tenderly, in ancient Persia a poet wrote of the battle between two noblemen. "They fight day and night to decide which army shall do a twenty on the other's ninety." The hand sign for twenty placed the thumb of the right hand between the index and third fingers, so that it protruded (a relative of today's erect middle finger). The ninety, on the right hand, involved curling the index finger so that its tip touched the bottom of the thumb, leaving a small round hole (not the closed virgin's crown, but now open, perhaps an anus) between the two fingers. Doing the twenty on the ninety wasn't a nice thing to say, even for a poet.

## AND THEN THE BODY

Body counting was even more complex. It began with the fingers and moved right up the arms, eventually to number the whole body. The Paiela, who lived in Papua New Guinea, started with the little finger of the right hand for the number one, and then followed the fingers, wrist, elbow, shoulder, and ear before going across the face, including eyes, nose, and mouth, and coming down the other arm to the little finger—twenty-two—on the left hand. The word for each of those numbers was the same as the name of the body part. To name a number, you could just point, or you could use a word in a sequence, so it was clear that you were talking about thirteen, and not just pointing at your mouth. By the same token, it was possible to count without saying a word, just pointing at one thing after another. Counting, then, becomes a matter of gesture.

There were the usual variations. In Paraguay, the Lengua had special words for one and two; three was *made-of-one-and-two;* four was *both-sides-same.* After four came *one-hand* for five, *reached-other hand-one* for six, and ten, *finished-both-hands.* Onward, with the same kind of phrasing, for the feet. Twenty, then, was *finished-both-feet.*

The Zuni, in Africa, did something similar, but with a touch of poetry. *Taken-to-begin*—the first finger—meant one; *raised-with-the-previous,* two. *The-finger-that-divides-equally* was three, and *all-fingers-bar-one,* four. After that, things were more sedate.

Georges Ifrah, in *The Universal History of Numbers,* writes about how tribal shamans could predict important dates. First, the shaman would announce the arrival of the next full moon (sunrise and sunset counted the days, but the cycles of the moon made it possible to count the future). Using both words and gestures, the shaman might say that a festival would fall on the thirteenth day of the eighth month to follow that next full moon. "The moon that has just risen must first wax and wane completely," Ifrah imagines the shaman saying. "Then it must wax as many times again as there are from the little finger on my right hand to the elbow on the same side. Then the sun will rise and set as many times as there are from the little finger on my right hand to the mouth. That is when the sun will next rise on the day of our Great Festival."

The tribal chief, just to be sure—not because he doubted the shaman, but just to have a record—would daub paint on the relevant parts of his body (a circle for the moon's phases, a thin line for the sunrises and sunsets, and a thick line on one eye to mark the day that the festival would begin). The chief, in effect, became the walking embodiment of a wall calendar. Someone else in the tribe would keep track of the days of the lunar cycle on a tally stick, making grooves in the stick for the days and tying a string around each groove as that day passed—as we put a large X in a calendar box to mark that that day is over.

## ON FIFTH BASE

Today we count by tens. That may not be the most practical way to count—ten can by divided by only five and two and one, whereas twelve, say, can be divided by one, two, three, four, and six—but it's awfully popular even so, because ten is at our fingertips.

It would be easy to say that where it's very cold, people might want to have only one hand at a time out in the weather, and so in those places, people count by fives. Perhaps. But there are many places where it's the rule to count by fives, and the weather doesn't usually have anything to do with it. One hand counts; the other points.

> **There are only five things you can do in baseball—run, throw, catch, hit, and hit with power.**
>
> **LEO DUROCHER,** FORMER NEW YORK GIANTS MANAGER

A language in the New Hebrides has words for numbers up to only five; everything after that is a compound, just as our -teen numbers combine with those that have come before, and our numbers that mark tens (twenty, thirty, and so on) combine the -ty with numbers from one to nine before switching to the next higher group. Languages everywhere from the Caribbean to Africa to Asia show similar traces for counting by fives.

The Romans, with their stately numerals, counted by fives. They chose V to stand for five—perhaps imitating the image of the gap between the thumb and the four other fingers on a single hand and indicating thus the whole hand. X was ten—two hands, a V on the top and another inverted on the bottom.

Roman numbers go up by fives: V, X, XV, XX, XXV, or five, ten, fifteen, twenty, twenty-five, and so on. Ones are added before or after each five being counted, to show the numbers between: IV is four; VI is six.

As the symbols go higher, they're multiplied by five and doubled by turns. Five times I is V. V doubled is X, ten. And then, in turn, X multiplied by five is the next new sign, L, fifty. L is doubled to make C, hundred. Then C multiplied by five is D, five hundred. And D doubled makes M, one thousand.

There were symbols for numbers over one thousand, and they evolved over time, but they weren't used or needed very often, perhaps only to count soldiers, or to indicate the amount of an inheritance or any other large sum of money—income from a province, say.

It wasn't originally the Romans, by the way, who wrote their numbers by putting I before another symbol to indicate one less than that number. They wrote IIII for four, or VIIII for nine. The new system, the one we use, was begun later because it made for shorter and clearer numbers. V and X now follow the old rule of four, that it's possible to count up to only four at a glance. IV is much easier to read than IIII, and after all, four comes before five. Nine, instead of being VIIII, became IX.

Even so, Roman numerals tend to be long and confusing. The higher they go, the more space they stretch across. CMXLVIII is harder to write and to read than 948. You need to know what each symbol means, and you have to be able to add from left to right: C is 100; CM is 100 less than 1,000, or 900; plus 10 less than 50, which makes 940; plus 5 and 3—aha! 948. The numeral 948 is faster. You have to stop and figure Roman numerals out, and hope that you can still remember whether it's D or L that means 50. As for arithmetic—adding and subtracting, multiplying and dividing—forget about it. Roman numerals look grand on monuments. They don't do much on flash cards. If anybody still uses flash cards.

Roman numerals are still used. Copyright dates, especially in films, are often in Roman numerals, as are the numbers for introductory pages of books. Some watches and clocks use Roman numerals—sometimes even using IIII for four and VIIII for nine. The Olympics are counted in Roman (not Greek!) numerals, and so are American football's Super Bowls, though that didn't happen until the fifth year—the numerals were then applied retrospectively to the first four Super Bowls. Writing in the *New York Times,* John Branch says, "Certainly, the confusion of keeping the games straight, particularly in a sport in which the regular season is played in one calendar year and the Super Bowl in another, was one reason for the denotation." He also notes that the Roman numerals lend the players a bit of the gladiator image that "popular culture has attached to football players." The fact that using Roman numerals gives the games a bit of grandeur probably didn't hurt either.

> A lot of people probably think the N.F.L. [National Football League] invented Roman numerals.
>
> **BARRY JANOFF,**
> EXECUTIVE EDITOR,
> *BRANDWEEK* MAGAZINE

Names of kings and sons—and sons of sons—have Roman numerals stuck after them: James III sounds more important than James No. 3. (For sons, the first is just junior, not II (II has the same name as his grandfather); the next is III.) Legal documents often use Roman numerals, and some periodicals count their editions that way—vol. VII, no. 6. And there are two world wars counted that way, I and II, and the hope is that there will never be a III.

## LOOKING AT FIVE

Where did Roman numerals come from? Perhaps from tally sticks—those series of notches that marked early counting. Making a series of cuts without any differentiation would have made keeping track very difficult. Once four notches had been made, making the fifth slightly different kept the counting in groups, just like our hands. It was the rule of four again: four could be counted at a glance; more could not, so the fifth notch had to be different. It was common to make it with an added slant—at a glance, each five mark looked like a hand, with four fingers upright and the thumb at a slant. That mark turned into a V.

> I never put on a pair of shoes until I've worn them at least five years.
>
> SAMUEL GOLDWYN

After five, the same thing happened until the tenth notch was reached, when, most often, the slanted upright was crossed to make an X. Going higher, XV became fifteen—which was one hand after the first two—and so on. New marks were needed for fifty and a hundred, but for the most part, the notches became a combination of cuts already made. The notches could be transcribed—to a board or a counting table—by copying how the tally stick looked. But instead of marking IIIIV, like the tally stick's notches, now all that was needed was the V—that meant five; and instead of IIIIVIIIIX, all that was needed was just the X. (The other numerals evolved from various lines and curves, but C was also the first letter of the Latin *centum,* "hundred," and M was the first letter of *mille,* "thousand.")

We make our own bundles of five using the same principle when we make four straight lines on paper—or on the jailhouse wall—and then add a slanted line across all four to make a little bundle of five. We repeat the bundle for ten, instead of making a different mark when we reach that number. We're still counting on our fingers, as it were, but on paper now, still in groups of five, just as the Romans did, and the Venerable Bede, and all those who came before and after.

Thinking in fives makes it possible to multiply and divide on our fingers as well as to add and subtract. In parts of France, it was long the custom to work numbers this way. How was it done? Suppose you're multiplying nine by eight. Consider how many above five each of those numbers is. On your left hand, bend down four fingers—because nine is four more than five. On your right hand, bend down three fingers, because eight is three more than five. Count how many fingers you've bent down altogether: seven. This is the number for your tens column: you're going to be in the seventies. Now multiply the unbent fingers on one hand by the unbent fingers on the other: one times two. This is the number for the ones column. Answer: seventy-two.

The written numeral 5 didn't evolve quite as logically as the numbers that came before. For a while, it looked like a curvy sort of Y; then like an upside-down h, then an h on its side, a 3 facing the other way, and gradually, something that looked like an S, and at last an S with its top curves changed into straight lines. A sideways cup with part of a box on the top. It could hold the world.

The word *five* has a double set of parents: the Sanskrit *panca* to the Greek *pente,* and the Latin *quinque* through the Spanish *cinco* and the French *cinq. Five* is from the Old English *fif,* from the

Anglo-Saxon *fife,* from the Old High German *finf,* or *fünf. Fünf* and *five* are cognates, so closely related that it would probably be against the law for them to marry and have a baby named Ten.

A pentagon is a figure with five sides—just like the one in Washington, D.C., but almost always smaller. Pentameter is poetry with five repeating feet (iambs) per measure; Shakespeare wrote much of his poetry and drama in iambic pentameter. (Iambic pentameter uses an unstressed syllable followed by a stressed syllable: da *dah.* A line of iambic pentameter is five da-*dah*s in a row.)

> A hero is no braver than an ordinary man, but he is braver five minutes longer.
>
> RALPH WALDO EMERSON

*Quinque* gives us quintet, a group of five, and quintuplets, who happen more often than they used to, though none will ever be more famous than the Dionnes, if you're of a certain age. They were born in 1934, and were a sensation, because they were the first quintuplets to survive infancy.

Five is close to our hearts—though our hearts have four chambers, not five. In addition to our five fingers, we have five senses, taste, smell, hearing, sight, and touch—leaving the sixth for extrasensory perception. Our taste buds reduce the complication of eating to five basic taste sensations: sweet, salty, sour, bitter, and umami—and any number of combinations. Umami isn't well defined, but it's a kind of savory flavor—like that of Parmigiano Reggiano cheese, or walnuts.

If you're near a telephone, feel the button for the number 5 (which is also JKL). It has a raised dot in the middle. Anyone who is partially or completely blind can dial by feeling the 5 button; all the other numbers are arranged around it. There's the same dot on the 5 on most computer number pads; if you can find the 5, with its raised dot, you can find them all. Same thing is true on keyboards: There's a raised dot on the F and another on the J. If you know touch-typing, that's where you align your fingers.

There are five vowels in English: a, e, i, o, and u. (There's a word that includes all five vowels in their proper order: *abstemious.*) You can say vowels without moving your lips or your tongue, making *Mama* the first word babies say in almost every language, because it's the easiest word to say: all you have to do is open your mouth twice. Or once, for *Ma. Papa* is a little harder, involving as it does a tiny puff of air, and *Dada* is harder again, because although the mouth doesn't have to move to say it (assuming it's already open), the tongue does.

Most of us work five days a week, and when we want to rest, we take five. (Some say that the expression "take five" began because that was the time it took to smoke one cigarette on a musician's work break.) There are five oceans: Atlantic, Pacific, Indian, Arctic, and Southern. Five o'clock shadow is the darkness that overtakes a man's face toward the end of the afternoon, a long time since the morning shave. "Gimme five," a slightly dated expression, asks for the open-handed (palm-up) salute; high fives are higher up.

Less open is the fifth column, people working inside one country to support an enemy country. The expression was first used in 1936, during the Spanish Civil War, by a Spanish Nationalist general who said that as four of his army columns were moving to attack Madrid, his fifth column was working inside the city, to undermine the government from within. During World War II, German minority organizations in Poland and Czechoslovakia formed fifth column groups to help the Third Reich in its invasion of those countries. The American government justified its illegal internment of Japanese Americans during World War II by saying that they would otherwise act as

> It is only necessary to make war with five things: with the maladies of the body, the ignorances of the mind, the passions of the body, the seditions of the city, and the discords of families.
>
> PYTHAGORAS

a fifth column. There's no end to it: in France during the time of the Dreyfus Affair, the loyalty of French Jews was suspect; Irish Catholics living in England were seen as a fifth column during "the Troubles"; and today, there are those in Europe and the United States who are afraid of local Muslims in exactly the same way. Somebody else is always "the other," different from us, frightening to us. The fifth column—if it's to be effective—has to look the same as everybody else, but it is difference that we see as ominous.

Five-star, on the other hand, means the best—a five-star hotel is of the highest quality, and a five-star general is the army's highest-ranking officer. The star itself has five branches. It's a pentagram.

The pentagram is a five-pointed star, and inside the lines that lead to the star's points is another pentagram, not a star, just a five-sided figure. As a symbol, the pentagram has had great significance to a host of cultures, in South America, India, China, Greece, Egypt. . . . It was scratched on the walls of Stone Age caves; it's in Babylonian drawings and Sumerian writings from 3000 BC; and there are many references in the Bible to pentagrams. Christians once believed that the pentagram represented the five wounds of Jesus, but today pentagrams are much more often associated with Neo-pagans and witches and Satan worship. For Wiccans, the pentagram has long been a symbol of faith—as is the cross for Christians, the Star of David for Jews, and the crescent and star for Muslims.

For the early Christian Gnostics, the pentagram was a symbol of Venus—the female principal, the secret goddess. As a female symbol, the pentagram was embodied by a five-petaled rose. In *The Da Vinci Code,* Dan Brown writes that "the Rose was a symbol that spoke of the Grail on many levels—secrecy, womanhood, and guidance—the feminine chalice and guiding star that led to secret truth." That magic flower can still be seen as part of the ornamentation of many Gothic cathedrals.

The pentagram almost became the official symbol of the first Olympic Games—to be replaced at the last minute by five intersecting rings because it was decided that the circles were more representative of the Olympic spirit of inclusion and harmony. The five rings now represent the inhabited continents (excluding Antarctica, and counting North and South America as one) from which the Olympic athletes come.

Pythagoras believed that the number five represented man because of the fivefold division of the body, and held that the five points of the star represented the five elements that make up man: fire, water, air, earth, and spirit. The pentagram was sacred to the Greek goddess of healing; her name—Hygeia—was thought to be an anagram in Greek for those five elements.

## FIVE MAGIC

Five has, as so many numbers do, religious and mystical significance far beyond simple counting. Five is particularly important in Islam. The Five Pillars of Islam are the foundation of Muslim life: belief in the oneness of God and the prophethood of Muhammad; daily prayers; concern and charity for the needy; self-purification through fasting; and the pilgrimage to Mecca for those who are able. There are five daily calls to prayer and five fundamental elements of the pilgrimage to Mecca.

A five-fingered hand—three upraised middle fingers with two symmetrical smaller fingers on either side—is an important symbol for both Muslims and Jews. In Islam, the Hamsa (literally "fivefold") is the Hand of Fatima. Fatima was the daughter of the Prophet Muhammad and the wife of Ali, the Prophet's nephew. Miracles are attributed to her, and she's known as a faithful, holy woman. The Hand of Fatima is a symbol of patience, abundance, and faithfulness. For Jews, the hand, called the Hamesh ("five") represents the books of the Torah—the first five books of the Old Testament—and it's called the Hand of Miriam, who was the sister of Moses. It offers protection and good luck. There are Middle Eastern peace activists who wear the stylized hand symbol to represent the connections—the origins and traditions—that exist between Muslims and Jews.

For Christians, Pentecost (in Britain, called Whitsunday) celebrates the descent of the Holy Spirit upon the Apostles, fifty days after Easter. The word *Pentecost* comes from the Greek *pentekoste*—fiftieth day.

> **Five is the soul of man. As man is composed of good and evil mixed, so is the Five, the first number holding odd and even.**
>
> SCHILLER, *DIE PICCOLOMINI*

The first five books of the Old Testament together form the Pentateuch: Genesis is the story of creation and the Garden of Eden, and ends with the clan of Jacob leaving Canaan, because of famine, for Egypt. Exodus is the story of Moses leading the Jews out of slavery in Pharaoh's Egypt. Leviticus establishes basic laws of morality and conduct. Numbers includes the wandering of the Israelites in the desert, and ends at the brink of the promised land. In Deuteronomy, Moses sees the promised land from a mountain, but dies and is buried before the Israelites enter Canaan again. Traditionally, the Pentateuch was believed to have been written down by Moses; the five books form a continuous story, from the Garden of Eden to Canaan.

The ancient Chinese counted five elements: wood, fire, earth, water, and metal. Fire is pure yang; water, pure yin; metal and wood are both mixtures of the two, and together those elements and their mixtures form Earth. The arching curves of yin and yang also represent the sun and the moon, and the five elements represent the five planets known to the ancient world.

Buddhist mandalas, designs used in healing, initiation, meditation, and transformation rites and for gaining wisdom and compassion, are often in a design called a quincunx: an arrangement of five objects in rectangles, squares, or circles, with one thing at each of the four corners and one at the center—like the symbols of suits on a five in a deck of cards, or the dots on a domino for the number five. (The word *quincunx* comes from the Latin for "five" (*quinque*) and for a "twelfth" (*uncia*) and had originally to do with a Roman coin that weighed twelve ounces. A quincunx pattern was also the standard formation for the elements of a Roman legion.)

The east–west line of the Tibetan Buddhist quincunx is the connection of mind, while the north–south line is that of action, or involvement in the "real" world. The two lines cross in the middle, and the intersection is the place of universal mind and energy, called the fivefold heart. The Navajo have the same crossing of the real and sacred worlds. (Like Tibetan Buddhists, they also create paintings with sand.) They call the center crossing the meeting of the Holy Wind and the Universal Mind, the place where all things originate.

For the Chinese, the importance of five as a quincunx goes back many thousand years, to a pattern said to show the laws of world order. In its center is a quincunx of five dots, the balancing force and unifying field of the pattern and of the universe as well.

Five is the number of letters in the words *alpha* and *omega,* Greek letters which are the beginning and the end. All Chinese music is based on a scale of five tones.

For the Japanese, five frames the verse form known as haiku, poems written in three lines, traditionally with a total of seventeen syllables, and with the lines having a syllable pattern of five-seven-five. Limericks have five lines, and they're usually jokes. They have a rigid rhyme scheme—aabba—and a fairly rigid rhythm. There was a young lady from somewhere. . . .

Haiku offer a moment in time, a focus on one small thing, but with a glimpse of the whole, the continuity surrounding *now* with all that comes before and after. Haiku go to the quintessence, the essential nature, of a thing by describing only a bit of the whole, Quintessence is another five word, and a particularly lovely one. To the ancient Greek belief that there are four elements (earth, air, fire, and water), the Pythagoreans added a fifth, a fifth essence—the nether, or ether, which they said flew upward at the moment of creation and became the substance out of which the stars were made. They believed this quintessence to be latent in all things. Eventually quintessence came to mean the highest nature, the most concentrated form, the most typical example of the thing itself, the condensation of itself, the part beyond, above, almost distilled from, the rest.

> Let there be light, said God; and forthwith light Eternal, first of things, quintessence pure, sprung from the deep.
>
> MILTON

The Pythagoreans also named five as the number for marriage because they believed two to be the first even number and three to be the first odd number (since one was not considered a number). Two and three—even and odd, female and male—make five. And after the marriage, five was associated with children (the hypotenuse of a Pythagorean triangle was numbered five and was considered the offspring of the other two sides) and with a state of nature. Indeed, nature builds on five-pointed stars and five-petaled flowers and limbs that end in five fingers or five toes.

For all of that, for all of its importance, the quintessence of our hands, five is still a number that somehow feels not quite finished.

It wants to get to ten.

# 6

## SIX IS PERFECT

Now we need two hands.

Six is an ongoing sort of number; what one hand has finished, the next begins. It's a small leap from five to six—over the body and across the mind—but it's a logical and inevitable one. Even alone there on the other hand, six has a feeling of connection. More than that, it has a feeling of commitment: the count goes on!

> **Life begins at six—at least in the minds of six-year-olds.**
>
> **STELLA CHESS, PSYCHIATRIST, AND JANE WHITBREAD, WRITER, IN *DAUGHTERS***

Mathematically, six is not just an afterthought to five; it's a number of distinction. Six is perfect. That doesn't mean it's beautifully made—though it is, with its graceful loop and its long tail, a perfect little sperm, albeit swimming downstream (nine goes in the proper direction). Being perfect also does not mean that six is full of magic and mystery, though it has its moments—one of them on the forehead of the beast (see below). What makes six perfect is simple, if you keep your head about you. A mathematically perfect number is one that is equal to the sum of all the numbers that divide into it except itself. It's equal to the sum of its parts. Six can be divided by one, two, and three, and if you add those numbers, the sum is six. Perfection.

The next perfect number after 6 is 28, which can be divided

by 1, 2, 4, 7, and 14, adding up to 28. It takes a while longer to reach the next perfect number, which is 496, and the one after that is 8,128. The next one is in the millions (33,550,336).

The first four were known by the first century. The next one was found fourteen centuries later. A note: all the perfect numbers (so far) are even numbers. And the last perfect number found (though there may be perfect numbers missing between those we know about) has over eighteen million digits. It was found in 2005.

The Pythagoreans found the first two perfect numbers, 6 and 28, early on; from this, they developed number theory, the study of numbers and their characteristics. The Greeks used pebbles to arrange and represent numbers, and noted that 3, 6, and 10 can be arranged as triangles: For three, one pebble centered above two makes a small triangle. For six, there are three pebbles in the bottom row, two in the middle, and one on the top. For ten, the rows are four, three, two, and one. They also noted that adding consecutive numbers equals their triangular numbers: 1+2=3; 1+2+3=6; and 1+2+3+4=10. They also made numbers that formed squares (4, 9, and 16) by adding consecutive *odd* numbers. Two rows of two pebbles is the four square (1+3=4). Three rows of three pebbles makes the nine square (1+3+5=9). Four rows of four pebbles makes the sixteen square (1+3+5+7=16), and so on. But only six is perfect.

> A good poet is someone who manages, in a lifetime of standing out in thunderstorms, to be struck by lightning five or six times.
>
> RANDALL JARRELL

A final note on perfect numbers. Most of nature's creatures have four legs—cats, dogs, tigers and lions, giraffes, elephants. No matter how many legs, though, everything that has legs has an even number. Birds and people have two, horses and cows have four, spiders have eight. Even centipedes have two columns of legs, with the same number in each column, for an even total. But flies have six. What a comeuppance for a perfect number! To be chosen by a bunch of yukky flies!

## THE SHAPE OF SIX

In Greek, six was *hex,* and so a hexagram is a shape with six sides. One of the best-known hexagrams is the Star of David—the Magen David ("the shield of David"). The star is made up of two intertwined triangles, inverted over each other to form a six-pointed star. It's used on the Israeli flag, and is a symbol of Judaism generally.

The star has been used as a symbol of religious identity—as the cross is for Christians—for centuries, and it has also been used by many other groups over the years. Some Baptists believe that God made a covenant with the Israelites giving them their land, and so some Baptists use the Star of David in their churches and stained-glass windows as a symbol of that covenant. There is a Star of David embedded in the ceiling of the National Cathedral in Washington, D.C., and four more in various windows. The Magen David is used in the emblem of the Theosophical Society; it's worn as a badge by members of the Zion Christian Church in Africa; it's used by the Church of Jesus Christ of Latter-Day Saints as a symbol of the tribes of Israel and of friendship for Jews.

In the Koran, it is written that David was a prophet and a king; he is among the First Testament figures who are revered by Muslims. His star is seen in mosques and on Islamic and Arabic artifacts. Six-pointed stars are also found in Hinduism, Buddhism, and Jainism. In Sanskrit, the two triangles are called *Om* and *Hrim;* together they stand for man's position between earth and sky. The downward triangle is a symbol for Shakti and the upward triangle for Shiva. The union of the two represents Creation.

For Jews, one of the two triangles points upward toward God; the other, downward toward life in the world; because they're intertwined, the two are inseparable, as are the Jewish people.

Probably the most infamous use of the Star of David was the yellow patch which Nazis required Jews over the age of six to wear in the German-occupied areas of Europe during the Holocaust. Each country had its own version: the German patch had the word *Jude*—German for "Jew"—written across its middle; in France, the word was *Juif;* in Holland, *Jood.* In Poland, Jews had to wear not only the patch, but also a white armband with a blue Star of David on it—and a second yellow patch on their backs so they could be seen coming and going as well as sideways. Even in concentration camps, Jews had to wear yellow patches.

> Speeches are for the younger men who are going places. And I'm not going any place except six feet under the floor of that little chapel adjoining the museum and library at Abilene.
>
> **PRESIDENT DWIGHT D. EISENHOWER**
> ON THE EVE OF HIS SEVENTY-FIFTH BIRTHDAY

After the war, the star became the proud symbol of Jews around the world, and it is in the center of the flag of the state of Israel. The Magen David Adom (Red Star of David) is Israel's official emergency service, like the Red Cross in the United States and the Red Crescent in Islamic countries.

If there is a link between the Nazis, Adolf Hitler, and the number six, it may very well be the triple six which is 666, the Number of the Beast described in the New Testament book of Revelation (13:18). The King James version says, "Here is wisdom. Let him that hath understanding count the number of the beast; for it is the number of a man; and his number is Six hundred three score and six." Some say the Beast is Satan, or the Antichrist, who has the number 666 written across his forehead or on his right hand.

If you use a code in which the letter A stands for 100, B for

101, C for 102, and so forth, the letters in Hitler's name add up to 666 (107+108+119+111+104+117=666). If starting with 100 seems too contrived, it's possible to get the same 666 by using 1 for A and 8 for H (as in Adolf Hitler), and decode his initials as 18, which is three sixes (six three times equals eighteen), or 666.

That kind of mathematical contrivance has been used over and over to point to possible Beasts. One of the first to be called the Antichrist was the Roman emperor Nero. If one uses the Hebrew letters that spell the words *Caesar Nero* as numbers, their total is 666. Nero had been responsible, among his many other crimes, for considerable persecution of Christians. In *The White Goddess,* Robert Graves wrote about the emperor Domitian, another persecutor of Christians. Graves noted that the Roman numerals in descending order from 500 to 1 (D, 500; C, 100; L, 50; X, 10; V, 5; and I, 1) total 666. He noted further that the numerals are an acronym for the Latin sentence *Domitianus Caesar Legatos Xti Violenter Interfecit*—or "The emperor Domitian violently killed the envoys of Christ."

After Martin Luther had been excommunicated in 1521, a Catholic theologian said he was the Antichrist, because Luther's name in totaled Roman numerals gives the number 666. Luther's disciples found a way to respond. They took the letters in the phrase *Vicarius Filii Dei* (Vicar of the Son of God), which is on the papal tiara, and showed that they, too, added up to 666.

Number and letter tricks aside, 666 still reverberates as something to be dreaded. It's considered "the Devil's number," and it has its own phobia: *Hexakosioihexekontahexaphobia,* fear of the number 666—not unlike *triskaidekaphobia,* fear of the number 13, but just that much scarier. Residents of Colorado, New Mexico, and Utah petitioned the federal government to have the number of U.S. Highway 666 changed. Perhaps it was paved with good intentions? A spokesman for the New Mexico Department of Transportation noted that once the state decided to go ahead and change

> Alice laughed. . . . "One can't believe impossible things."
> "I daresay you haven't had much practice," said the Queen. "When I was younger, I always did it for half an hour a day. Why, sometimes I've believed as many as six impossible things before breakfast."
>
> LEWIS CARROLL, *ALICE IN WONDERLAND*

the highway's number, all of the signs labeling it as U.S. 666 were stolen. The sheared metal stubs that had held the signs were all that was left on the two-lane road known as "the Devil's Highway." For a while after the number change in 2003, eBay listed at least three Route 666 signs—and hundreds of T-shirts. Route 666 is now an innocuous but safe Route 491.

The United States is not alone in its feelings about 666. South Korea added seven soldiers to its original contingent sent to Iraq to make a revised total of 673. There was a bus route in Moscow that had the number 666; in 1999, it was changed to 616. There's still an A666 in England; its nickname, predictably, is also "the Devil's Highway." Tijuana, in Mexico—known as the Mexican Sin City—had an area code of 666 until, after considerable local protest, it was changed to 664. Japanese coins number 1, 5, 10, 50, 100, and 500 yen. You know the total, and you know the reaction. A remake of the 1976 movie about the Devil, *The Omen 666,* began production in Croatia, but filming had to be moved to the Czech Republic and Italy after protests by Croatian authorities and local churches. The movie opened on—ta da—June 6, 2006 (6/6/06).

The numeral 666 has had other movie moments. In Stanley Kubrick's *A Clockwork Orange,* Malcolm McDowell's character was a stand-in for the Beast. In one scene, he's shown between two police officers; their numbers are 665 and 667. Quentin Tarantino, in *Pulp Fiction,* shows a briefcase that's opened by the code 666. The briefcase belongs to Marcellus Wallace, a soulless and evil man played by Ving Rhames. *The Phantom of the Opera,* the Andrew Lloyd Webber musical, opens with an auction; the lot number of the restored chandelier is 666.

Then there's the worry, if your street address is 665 or 667, that you are "the Neighbor of the Beast." Not to fret—unless you live in number 666; that might be cause for nervousness. In many countries, 667 and 665 wouldn't be neighbors; they'd be across the street. It's 664 and 668 who should stay alert. The telephone number 1-800-666-6666 has been called "the Toll-Free Number of the Beast." $665.99 is "the Retail Price of the Beast," 00666 is "the Zip Code of the Beast," and the root of all evil is 666 inside a square root symbol. And 333 is the devil doing a halfhearted job.

> His words leap across rivers and mountains, but his thoughts are still only six inches long.
>
> E. B. WHITE, "WORLD GOVERNMENT AND PEACE," *THE NEW YORKER*

## MAGIC NUMBERS

Numbers have been used like this for centuries—to add up to some kind of significance, one way or another. Numbers have always had a dual nature, practical on the one hand—*How many? How much?*—and mystical on the other. It's only recently that these two realms, the utile and the magical, have been separated. For centuries, they were seen as mutually reinforcing—even more, as inseparable.

Numbers, after all, are a guide of sorts to spiritual knowledge. The Bible is full of numbers: the Israelites wandered in the desert for forty years; there are twelve tribes of Israel and ten of them are lost; it rained for forty days while Noah was building the ark, into which he led the animals two by two. The Trinity is the personification of three; there are four evangelists; Jesus had twelve disciples;

the Antichrist can be identified by the number on his forehead. The book of Revelation is rife with the number seven. All of that is just the merest scratching of the spiritual and numerical surface.

The Phoenicians, the Hebrews, and the Greeks all used letters to double as numbers. In Greek, the letter alpha, A, also literally means 1. In Hebrew, aleph, still A, again means 1. The first letter doubles as the first number. Out of this confusion of letters and numbers grew other kinds of numerical mysticism: the Jewish Gematria (a word possibly related to the Greek *geometrikos arithmos,* "geometrical number"), the Muslim *khisab al jumal* ("calculating the total"), and the Greek *isopsephy* (again, "adding the numbers").

When letters in those alphabets were put together to spell a word, the word then had a dual meaning—and one meaning was based on the arithmetic total of its numbers. Two words with the same total were considered equivalent, making for definitions that had nothing to do with the literal meaning of the words, but had much to do with their newly discovered mystical meanings. Gematria was used to interpret the Bible (and there are those who believe it may even have influenced the writing of some of its passages); it was used in Greek mythology and literature; eventually it was used to reinterpret the past and to predict the future; and finally it descended to the deciphering of personality. Today, numerology is based on this same idea: if you give a numerical value to letters, you can tell what underlies the meanings of the words they spell.

A sweet example of Gematria is based on the Hebrew words for wine and for secret. Given their numerical values, both words total 70, and therefore have an equivalency. Some rabbis combined the two words in the phrase *nichnas yayin yatsa sod*—the key words are *yayin,* "wine," and *sod,* "secret"—or "the secret comes out of the wine," better known in Latin as *in vino veritas,* "the truth is in the wine." When you drink, you tell.

More religiously, the Hebrew words for love (*ahavah*) and one (*ekhad*) both total 13. In the terms of Gematria, this corresponds to the belief that God is Love: "One" represents the credo that there is one God; and "Love" stands for the force which is the basis of the universe (as in Deuteronomy and Leviticus). The sum of the two words is 26, which is the numeric total of *Yahweh,* considered the true name of God—so holy that it is not to be spoken or written casually; it is usually read as *Adonai,* "My Lord." The common Semitic word for God is *El,* as in Israel; the Torah uses *Elohim* (a plural) to refer to God, but *Yahweh* incorporates the three Hebrew tenses of the verb "to be" (He was; He is; He shall be) and therefore speaks of the eternal nature of God.

There were those who let that same twenty-six reverberate much further. In Genesis 1:26, God says, "Let us make man in our image." Twenty-six generations, through the begats, separate Adam and Moses. And, if you like, the proof that God made Eve by using one of Adam's ribs can be found in the numerical difference between Adam's name in Hebrew, which totals forty-five, and Eve's, which comes to nineteen. The difference is twenty-six. Did you doubt?

In the New Testament, Jesus says that he is "the Alpha and the Omega." Alpha and omega are the first and last letters of the Greek alphabet; the meaning is thus that Jesus is "the beginning and the end of all things." In Matthew, the Holy Ghost appears at the birth of Jesus in the form of a dove; the Greek word for dove is *peristera,* and its letters add up to 801. The letters in the phrase "Alpha and Omega" also add up to 801. The phrase thus became a mystical affirmation of the Trinity, the Father, the Son, and the Holy Ghost.

What was true for Jewish, Greek, and Christian mystics was also true for Muslims. For them, though, a complication was added. Each letter of the alphabet also stood for a number—but in addition, each letter also incorporated an attribute of Allah in numeric terms. Alif, A, is the first letter of Allah; it is also the number one; it is also 66, the number total of the name Allah. Some other letter meanings and their numbers are Truth, 108; Light, 256; Eternal, 134; Judge, 65; and Guide, 20. These numbers are in addition to the numeric position of the letters in the alphabet, and to the numeric totals of words.

Out of the many numbers that resulted, forecasts could be made about a host of events—even the outcome of a battle—and the numbers could also be used to evaluate the past in new ways, deal with the present, and forecast the future. Some Muslims drew a magic square (more about magic squares when we reach nine), a good-luck piece that Georges Ifrah says not only offered the possibility of getting rich quickly, but also was a charm against evil and a way to bring upon themselves "every grace of God." The square—think of tic-tac-toe lines inside a frame—had 21, 26, and 19 in the top row of boxes; 20, 22, and 24 in the middle row; and 25, 18, and 23 in the bottom row. The total, to be added across and down each of the three columns as well as diagonally, is 66, the number of the name of Allah.

Part of the idea of magic numbers—for many different peoples at many different times, from Pythagoras to a numerological chart based on your name—originated from the feeling by contemporaries that the people who could work with numbers—count high and calculate—might be supernaturally skilled. Numbers were mysterious; few people understood them beyond what they needed to count what they had. For many years in Europe, arithmetic was cultivated by the priesthood, just as writing was, and the making of books, before Gutenberg came up with the godless printing press. The Greeks and the Muslims and the Jews made mysticism out of numbers. Numbers were part of religious rites and ceremonies; numbers were awesome and powerful. Just as astrology came before astronomy and alchemy preceded chemistry and religion came before science, numerology existed before there was a theory of numbers.

And now there's Arithmancy, an elective course at Harry Potter's Hogwarts School of Witchcraft and Wizardry. In Arithmancy, letters have values from one to nine. A is one, through I, which is nine; J is one again, through R, nine; S is one; and the alphabet stops at Z, which is eight. Adding the numerical value of the letters in a name (and the answers, if the answer has two digits) and then of the vowels and finally of the consonants gives three numbers: the Character Number (information about the personality), the Heart Number (the hidden fears and desires of the inner life), and the Social Number (the face shown to the world). The numbers in each category offer detailed interpretations of the person whose name is being studied—including possible good and bad days, and hidden kinships, good and bad, among people, places, and things.

Prisoners sometimes use numbers as a code for ideas. White supremacists, for instance, sometimes use 88 as a code symbol for Heil Hitler—the two H's standing for the double 8—a long numerical way from the 26 of "God is Love," or 66 as the name of Allah.

> Of course, behaviourism "works." So does torture. Give me a no-nonsense, down-to-earth behaviourist, a few drugs, and simple electrical appliances, and in six months I will have him reciting the Athanasian creed in public.
>
> W. H. AUDEN

## A SEXTET AT THE ORGY

The Greek word for six was *hex,* but Latin had it a bit differently: *sex*. Sex, of course, is a perfectly fine word in English, but it has nothing to do with six—unless you're at an orgy. Six came through a variety of turns: *se* in modern Irish, *saihs* in Gothic, *sehs* in Old Saxon, *sechs* in German, *chest* in Russian, and—strangely enough—*six* in French. A hexagram, on the Greek side, is an object with six sides; a sextet is a group of six. A semester once lasted six months; a sestet is six lines of poetry, often the last six lines of a sonnet, and a sestina is a poem of six verses (plus a final tercet) in which the last words of each line of the first stanza appear in a different order as the last words of the six lines of all the other stanzas. A sextant is a navigational instrument with a sixty-degree arc, or a sixth of a circle. A siesta is the Spanish sixth hour, midday in the old twelve-hour day that began at sunrise and ended at sunset—and a fine time for a nap.

Our modern numeral for six can be traced back to the Brahmans in India. They drew a six that looked like the script e, rotated sideways. Over time, the part above the loop became more curved, and the bottom part became straighter. When six reached the Arabs, they simply dropped the lower part, and it didn't take much time, after the numbers reached Europe, for the six to look as it does today.

In addition to being mathematically perfect, six is a good number, hardworking, full of earnestness and logic. It's proper. It's perfect. Goody-Six-shoes. But wait! There's more to six than shows on its hexagonal surface. After all, snowflakes, the essence—no, the quintessence—of nature's magic, little puffs of lace adrift in the winter sky, each one different from all of the others, each one separate and fragile, but together, joined to the others, overwhelmingly powerful, an avalanche of weight—snowflakes have six sides. That's because, prosaically enough, the molecules of ice line up in a hexagonal pattern.

> Gastronomical perfection can be reached in these combinations: one person dining alone, usually upon a couch or a hillside; two people, of no matter what sex or age, dining in a good restaurant; six people . . . dining in a good home.
>
> M. F. K. FISHER, *AN ALPHABET FOR GOURMETS*

So do the cells of bees, in their hives. These waxy cribs are somewhat easier to explain than the molecules of ice. Bees are no fools. In their apiarian way, they seem to have understood geometry, just as they've understood direction and day care, communication and air-conditioning. Their cells have to be contiguous, all touching each other, so that nothing can come between them, contaminating the honey or endangering the larvae. The cells also have to be regular. QED, they have to be either triangular, square, or hexagonal. Given that all the cells are constructed out of the same material, provided by the bees themselves, the hexagonal shape is the one that holds the most honey for the apiarian money, and therefore works the best. Charles Darwin called the cells of honeybees a masterpiece of engineering, "absolutely perfect in economising labour and wax."

For the Pythagoreans, somewhat later than the bees, six stood for something similar. They believed six to be a number that stood for—and provided—equilibrium (back to six's earnest perfection and prim logic?), and they symbolized it by drawing two triangles, touching base to base.

## A SIX-PACK OF SIXES AND THEN SOME

Ice hockey teams have six members, all ready to fight. At the other end of the cultural scale, a group of early-twentieth-century composers in France was known as *Les Six,* even though only four of them were really famous: Milhaud, Poulenc, Honegger, and Auric (plus Durey and Tailleferre). Their music was heard as a reaction to the heavy Germanic romanticism of Wagner and Richard Strauss, and to the lush impressionism of Claude Debussy, and they were lumped together as six individuals in one rebellion.

> Six of one, half dozen of another. Oh, well, half of one, six dozen of the other.
>
> **JOE GARAGIOLA**

Six comes in handy for a variety of adages—to be at sixes and sevens is to be in a state of disorder or confusion (it must be the seven that does it; six is perfect). Six of one and a half dozen of another has a lovely well-rounded sound, which my sister confused once by remarking that something was really six of one and seven of another. Well, yes. Could be.

There is the well-known six-pack, and a six-shooter is a revolver that can shoot six times after being loaded once. To deep-six something is to throw it overboard or, thus, to reject it or throw it away. It's from nautical slang for burial at sea—and it must also be an allusion to the depth of a grave on land: six feet under.

The *New York Post* has a gossip page called Page Six, which is where it originally was—and Page Six is a handy term for dishing the dirt. Henry the Eighth had six wives, and he wasn't very nice to most of them.

American senators are elected for a term of six years. There are six inhabited continents (Europe, Africa, North and South America, Asia, and Australia or Oceania). If you count Europe and Asia as one continent—Eurasia, since it's one landmass—there are still six continents if you add Antarctica. There are six sides to a cube, which is why six is the highest number you can throw with one die. And a frivolous note: This sentence would be seven words long if it were six words shorter.

There are six degrees of separation, the wonderful theory that anyone on earth can be connected to any other person through a chain of people that numbers no more than six steps. It is the very definition of "small world." It was first proposed in a short story in 1929. In the midfifties, two mathematicians began working on the idea and spent twenty years without determining whether or not it was true. Eventually, Stanley Milgram, a social psychologist, set up an experiment to test the theory, which he called the "Small World Theory." At random, he chose people in the Midwest to mail packages to people they didn't know, all of whom lived in Massachusetts. They had names and locations and were told to send the packages to friends who they thought would be most likely to know the addressee; the friends were to do the same again. Amazingly, it took an average of between five and seven steps—yes! six!—to get each package delivered. The findings were published in *Psychology Today,* and thus began the popular use of the phrase *six degrees of separation* (used by John Guare as the title of his 1990 play). In 2001, a professor at Columbia University moved the research forward another step, this time with an e-mail message as the package. Forty-eight thousand senders participated; their e-mails were addressed to nineteen people in 157 countries. The average number of steps from writer to reader? Six.

Six degrees of separation is the proud parent of the trivia game Six Degrees of Kevin Bacon. The goal of the game is to connect any film actor to Kevin Bacon in as few steps as possible. The method is to go through a series of movies that each actor was in, and by linking other performers with other movies, arrive at Kevin Bacon. Example: Take Elvis Presley. He was in *Change of Habit,* and so was Edward Asner. Edward Asner was in *JFK* with Kevin Bacon. Ergo: Elvis Presley has a Bacon number of two, which is very low.

## AND FINALLY, THE BEGINNING

More seriously, the world itself emerged in six steps. In the Old Testament, we read that God created the world in six days. He made light and darkness—day and night—on the first day; heaven on the second day; earth and its fruits, grasses, and seas on the third day; the sun and the moon and the stars—the firmament—on the fourth day; fish and whales and birds—the creatures of the sea and the sky—on the fifth day; the animals of the earth—including

Six is a number perfect in itself, and not because God created the world in six days; rather the contrary is true: God created the world in six days because this number is perfect, and it would remain perfect, even if the work of the six days did not exist.

**SAINT AUGUSTINE,**
*THE CITY OF GOD*

man—on the sixth day. His work done, in those first six days, says the book of Genesis, the first book of the Bible, God rested on the seventh day, the day after the six days of his labor, and blessed that day.

Six days were enough—perfect, in fact—but the day that was blessed was the seventh day.

# 7

## SEVEN IS A HAPPY NUMBER

Mathematically, seven stands alone. It's the only number in the first ten that is neither a multiple nor a divisor of any of the other numbers. One times two is two; two times two is four; three times two is six; three times three is nine; four times two is eight; and five times two is ten. That's all of them, and none of them is seven. Seven is a lonely number.

> Seven is a good handy figure in its way, picturesque, with a savour of the mythical; one might say that it is more filling to the spirit than a dull academic half-dozen.
>
> **THOMAS MANN,** *THE MAGIC MOUNTAIN*

Of the numbers from one to twelve—the only numbers before one hundred to have their own names—the only other solitary number is eleven. Seven and eleven are both also odd, indivisible—like three, the magic number—but also, of course, like five, all prime numbers, and also all defiant numbers, resolutely refusing to be divided into nice even piles of things. There's always something left over.

In its lonely way, seven seems to surround itself with mystery and potency. Even aside from Genesis (we'll get to that), the Bible is full of sevens. In Deuteronomy (7.1–2), God tells the Hebrews that when they reach Israel, they will displace seven nations. "When the Lord thy God shall bring thee into the land whither thou goest to possess it, and shall cast out many nations before thee, the Hittite, and the Girgashite, and the Amorite, and the Canaanite, and the Perizzite, and the Hivite, and the Jebusite, seven nations greater and mightier than thou." A lesson well learned; the covenant is the land of Israel and the people are the Jews.

For Christians, Revelation describes the Beast as having seven heads. The book of Revelation also refers to the seven churches of Asia Minor, seven stars, seven spirits before the throne of God, seven vials, seven horns, seven plagues, a seven-headed monster, the Lamb with seven eyes, seven seals (and the silence in heaven when the seventh seal is opened), seven trumpets blown, and seven bowls poured out. The number for the apocalypse is seven, despite 666 on the forehead of the Beast—the Antichrist whose arrival precedes the end of the world.

## SINS, VIRTUES, ANGELS, AND DEMONS

For the religious, there are seven deadly sins: pride, covetousness, lust, envy, gluttony, anger, and sloth. Have you forgotten any? There are also, for those of us who are good, seven virtues: faith, hope, charity, prudence, justice, fortitude, and temperance.

> One of the Seven was wont to say: "That laws were like cobwebs; where the small flies were caught, and the great brake through."
>
> FRANCIS BACON

There are also seven archdemons—one for each of the seven deadly sins—and seven archangels. The archdemons, the leaders of the demonic hosts, are Lucifer for pride, Mammon for avarice (or covetousness), Asmodeus for lechery (or lust), Leviathan for envy, Beelzebub for gluttony, Satan for anger, and Belphegor for sloth. Lucifer is a fallen angel, and he uses many names—including Satan, Beelzebub, Belial, and just plain old Devil. He's been an outcast since he was expelled from paradise, and a force of evil who is a thorn in the side of mankind.

The archangels are the good guys; they are God's messengers, mediators between man and God. They also command the heavenly army in the ongoing battle between Satan and the legion of angels. Michael is the leader of the heavenly host and a guardian of paradise. Gabriel is the heavenly messenger who told the Virgin Mary that she was to become the mother of Jesus, and he's also the trumpeter of the Last Judgment. Raphael is the guardian of human spirits. Uriel and Remiel are guardians of the underworld. Raguel is the avenger of God and the angel of earth; Sariel, the avenger of the spirit. (All of the names of the archangels end in *el,* Hebrew for "of God.") The Koran of Islam mentions four archangels and names two: Djibril (Gabriel), said to have revealed the Koran to Muhammad, and Mikhail.

There's a bit of coming and going among the archangels and the archdemons—this one is fallen; that one is redeemed—and there's also variation in the total number and the list of names, though Michael, Gabriel, and Raphael are on almost all lists. Archangels, you may already know, are next to angels in the hierarchy of angels. There are nine orders of angels, and we'll get to them when we reach nine.

The numbers seven, seventy, seventy thousand—and on into higher seventies—are used in Islamic tradition to mean a very high number, even infinity. It's considered that, of the first ten numbers, one is the smallest; two is just a step away from one; five and ten are exact numbers; four and six are just before and after five; and nine is only a step from ten. That leaves three, seven, and eight. Eight is close to the end of the ten numbers, and even aside from that, it's divisible by two. Three and seven remain, and three is the smaller, so seven is left, and thus it represents infinity.

There are seven earths and seven heavens in Islamic tradition, and Jews too have believed in a mystical concept of seven heavens. For Muslims, the farthest of the concentric spheres around earth contains the stars and is where God and the angels dwell—it is

> We shall never understand one another until we reduce the language to seven words.
>
> KAHLIL GIBRAN, "SAND AND FOAM"

Seventh Heaven. Kabbalists, Jewish mystics who find hidden insights in the Old Testament and in the traditional lore of rabbis, believed in a seventh heaven similar to that of the Muslims. For non-Muslims and non-Kabbalists, seventh heaven remains a place of bliss and joy—just that much better than cloud nine.

The magic and power of seven may very well go back to the ancient Sumerians, whose astrologers saw seven wanderers in the sky: the sun, the moon, Venus, Jupiter, Mars, Mercury, and Saturn. These, they believed, were the messengers of the gods, and much was named after them, and patterned according to them.

## SEVENS

In Sanskrit, seven was *sapta;* in Greek, *hepta;* in Latin, *septem;* in Old English, *seofon*—and thus, in English as we know it today, seven, clearly a weighty digit despite its slimness.

Hindus wrote the first sevens as a curved line with an arc at the top, sort of like an upside-down J. Gradually the lines straightened, and as it evolved, seven ran into the danger of being a six or nine look-alike, so the Arabs changed it into more of an upside-down V. It straightened again, and the next danger was confusion with the numeral for one. The first solution was to add a horizontal line above the number; and the second was to square the corner. In Europe, the horizontal line moved down to cross the stem of the seven, and today a discrete line across its middle remains to distinguish seven from one. In the United States, we let it all hang out: seven, plain, powerful, ominous, menacing. If you want to confuse it with one, that's okay with us. You take your chances.

For all of that, it's nice to know that seven is the smallest happy number, not counting one. Mathematically, that is, not just because it feels good to be seven. A happy number is one whose digits squared always end up equaling one. That demands further explanation: The square of seven is forty-nine ($7 \times 7 = 49$). So next, we square four and nine. Four squared is sixteen; nine squared is eighty-one. Next, $16 + 81 = 97$. Using that new sum, nine squared is eighty-one again; seven squared is forty-nine; and $81 + 49 = 130$. One squared is one; three squared is nine; and zero squared is zero; the total of one and nine and zero is ten. Ten breaks down into one and zero; one squared is one and zero squared is zero, and the total is now one. That's why seven is a happy number. And yes, there are also unhappy numbers, poor darlings. More of them, in fact. Only a few numbers—about 15 percent—are happy. But I think we knew that.

> Every action must be due to one or other of seven causes: chance, nature, compulsion, habit, reasoning, anger, or appetite.
>
> ARISTOTLE

## ONE HEPTAD AND ANOTHER

From the Greek for seven, *hepta,* we call a seven-sided figure a heptagon and a group of seven a heptad. You'd think that there aren't very many of them, because *heptad* is not a word you hear very often. Oh, but there are heptads upon heptads—hosts of heptads.

Rainbows, for instance, are made up of seven colors. Red is on the outside, then orange, yellow, green, blue, indigo, and violet on the inside. It's hard to distinguish the last inside colors—they're often lumped together as just blue or violet. The colors are a nearly continuous spectrum, and different people and cultures identify them differently. Even so, there are mnemonics for remembering the seven colors of the rainbow in sequence: "Richard Of York Gave Battle In Vain" (Richard III was defeated by Henry

Tudor at the Battle of Bosworth Field in 1485). In York, there's another version: "Rowntrees Of York Gave Best In Value." Or, if you prefer, you can remember the name Roy G. Biv. And then there's this: "Ring Out Your Granny's Boots In Vinegar." The last might make more sense if the boots were wrung rather than rung, but then red begins with r, not w.

Greek mythology had it that rainbows were a path made by a messenger between earth and heaven. Chinese myths described them as an opening in the sky that had been sealed by a goddess who used stones of different colors in her work. In Hindu mythology, a rainbow is the bow of Indra, the god of lightning and thunder. From Norse tales, it's a bridge connecting the homes of gods and humans. For Jews, according to the Old Testament, rainbows are a symbol of the covenant between God and man, and also represent God's promise to Noah that he would never again flood the earth. And for the Irish—remember *Finian's Rainbow*?—leprechauns have hidden a crock of gold at the end of the rainbow. Unfortunately, the end of a rainbow is a place that can never be reached.

Rome was famously built amid seven hills—Rome was supposed to have been founded by Romulus on the Palatine Hill. But Rome is not the only city built on seven hills; there are many others. Prominent among them are Jerusalem, Rio de Janeiro, Pretoria, Saint Paul, Seattle, Istanbul, Brussels, Prague, Bath, Cambridge, Athens, Lisbon, Bucharest, Moscow, and Melbourne. It seems natural to build in the bowl formed by the level area surrounded by hills, protected from the winds, like a saucer under the night sky. Why are there so often *seven* hills? Good question. But you could also ask, Why not? There are, we learn in school, seven continents and seven seas. The fabled seven seas of medieval and Arab literature probably meant the *many* seas, rather than that there were literally seven. The list of major seas changed from time to time, but usually included most of these: the Adriatic, Aegean,

Arabian, Black, Caspian, Mediterranean, and Red seas; the Indian Ocean; and the Persian Gulf. Today we can count seven oceans in our world: the North Pacific, South Pacific, North Atlantic, South Atlantic, Indian, Southern (or Antarctic), and Arctic. (If you combine North and South you won't have seven anymore. It does seem unfair to divide the Pacific and Atlantic in half and not even mention a variety of other considerable bodies of water.) The seven spikes in the crown of New York City's Statue of Liberty are meant to represent the seven seas and the seven continents.

> **By hard, honest labor, I've dug all the large words out of my vocabulary and shaved it down till the average is three and a half. . . . I never write "metropolis" for seven cents, because I can get the same money for "city."**
>
> **MARK TWAIN**

Looking up, we see two clusters of seven. Seven stars in the constellation Ursa Major, the Great Bear, are the principal stars of the Big Dipper. The Pleiades are another beautiful cluster of stars, this time in the constellation of Taurus. There are hundreds of stars in the Pleiades, but usually only seven are clearly visible to the naked eye. Folklore and mythology abound in stories about the Pleiades. According to Greek mythology the seven stars were originally the daughters of Atlas and Pleione. They all fell in love with gods and became the mothers of gods, and they themselves were changed into stars. The word *Pleiades* could derive from their mother's name, Pleione, or from the Greek word *plein,* "to sail," making the Pleiades the sailing ones, floating across the night sky. But there's also *pleos,* "full" or "many," which gives us the many stars, or *peleiades,* "a flock of doves."

In the Northern Hemisphere, the dawn rising of the Pleiades in spring and their setting in the fall mark the beginning and end of the seafaring and farming seasons. In South America, there are Indians who use the same word to mean both Pleiades and year; for them, the stars in their journey across the sky mark off the

> There are seven things of which you can never have too much: bread offered in kindness, meat of lamb, cool water, soft garments, beautiful fragrances, comfortable beds, and the view of everything that is beautiful.
>
> ARABIC PROVERB

beginnings and endings of the years. The Aztecs based their calendar on the Pleiades; across the Pacific, the Maori did the same thing; and Australian aborigines believed the stars formed the outline of a woman who had been nearly raped by the man who lives in the moon. Many Native Americans measured eyesight by the number of stars a viewer could see in the Pleiades cluster; so did some Europeans. To the Vikings, they were Freya's hens, and many old European languages compared them to a hen with chicks. In India, the Pleiades are called the Star of Fire, ruled by the Vedic god of the sacred fire. In Chinese constellations, they are considered to be the hairy head of the White Tiger of the West, and for Hindus, the name of the god Kartikeya means Him of the Pleiades. Finally, in Japan, the stars are known as Subaru, and they are the emblem of Subaru cars.

Seven sisters in the United States usually means something different—not the stars clustered in Taurus, the virile bull, but the private colleges founded originally to promote the excellent education of well-to-do women. They became the female Ivy League, and they are Barnard, Bryn Mawr, Mount Holyoke, Radcliffe, Smith, Vassar, and Wellesley. They weren't turned into stars, but they are now, for the most part, coed.

## AND MORE

There are many more heptads—we've barely scratched the seven surface. To start at the top, our heads have seven openings: two eyes, two ears, two nostrils, and a mouth. In ancient Egypt, carvers sometimes used a head to represent the number seven, and in India, where Hindus often used symbols instead of numbers (the moon as one, eyes as two), the head was a symbol for seven for the same reason.

Then there's the seven-year itch, the seventh wave, the lucky / unlucky seven in a game of craps (the dots on opposite sides of a die always add up to seven), and the lucky sevens many airplanes are numbered for: the 707, 727, 737, 747, 757, 767, and 777. So far.

> I was a good husband for seven years.
>
> JONATHAN SWIFT

Traditionally, there are seven arts and sciences; seven visible celestial bodies (the sun and the moon plus the five visible planets, according to ancient astrologers); the seven Graces; Seventh-Day Adventists; and the seventh-inning stretch. The visual displays on pocket calculators and digital watches have seven segments so that all the numbers can be shown; all the digits from 0 through 9 can be drawn with seven segments. Most people, asked to quickly choose a number between five and twelve, will say seven. There are seven notes in the musical scale, but strangely enough, we'll get to them when we reach eight.

Snow White had seven dwarfs as her dearest friends; Walt Disney called them Happy, Sneezy, Doc, Sleepy, Grumpy, Dopey—and nearly everybody forgets Bashful. They worked hard and loved her dearly but couldn't protect her from her wicked stepmother. They had to be away at work all day, while she swept and made up their tiny little beds, and combed her hair. She would be stupid about the apple; what could anyone do? Even seven was powerless.

> Because of Mozart, it's all over after the age of seven.
>
> WENDY WASSERSTEIN

Center fielder Mickey Mantle's number, when he played for the

New York Yankees, was 7, and one of the many famous episodes of the television series *Seinfeld* included George Costanza's devout wish to name his firstborn son Seven. He was engaged to Susan at the time, but poor Susan succumbed to a surfeit of poisonous glue while licking closed the invitations to her wedding, so Seven Costanza was never born—unless George got married after the series ended. Unlikely, alas.

Television also brought us *Gilligan's Island*—ninety-eight episodes, from start to finish—of the seven marooned on that remote and unlikely piece of land. As the song's lyrics told us, ninety-eight times, they were Gilligan, the Skipper, the Millionaire and his wife, the movie star, the Professor, and Mary Ann.

There were seven wonders in the ancient world (and none of them was Gilligan's Island): the Pyramids of Egypt, the Hanging Gardens of Babylon, the Statue of Zeus at Olympia, the Temple of Artemis at Ephesus, the Mausoleum at Halicarnassus, the Colossus of Rhodes, and the Lighthouse of Alexandria. Since most of the seven wonders no longer exist, there have been many suggestions for a new listing. Possibilities include the Roman Colosseum, Machu Picchu in Peru, the Great Wall of China, Angkor Wat in Cambodia, the Easter Island statues, Stonehenge, the Taj Mahal, the Parthenon, the Leaning Tower of Pisa, the carved city of Petra in Jordan . . . there are many more than seven.

Shakespeare wrote about the seven ages of man in *As You Like It:* "All the world's a stage, / And all the men and women merely players, / They have their exits and entrances, / And one man in his time plays many parts, / His acts being seven ages." They are the infant, the schoolboy, the lover, the soldier, the justice, the pantaloon ("with spectacles on nose, and pouch on side"), and finally, second childhood and "mere oblivion, / Sans teeth, sans eyes, sans taste, sans everything."

None of which, at any age, is to forget the man who was

traveling to St. Ives: "As I was going to St. Ives, / I met a man with seven wives. / Each wife had seven sacks, / Each sack had seven cats, / Each cat had seven kits. / Kits, cats, sacks, and wives, / How many were going to St. Ives?" It's an ancient riddle—it goes back to the time of the pharaohs. Whatever answer you get to the multiples of seven, don't forget to count the original man and the man he met. They were going to St. Ives too. (Unless you believe, as some do, that the man who was met was coming *from* St. Ives, in which case only "I" was going there. One.)

> I blame Rousseau, myself. "Man is born free," indeed. Man is not born free, he is born attached to his mother by a cord and is not capable of looking after himself for at least seven years (seventy in some cases).
>
> KATHARINE WHITEHORN, ENGLISH WRITER

The Chinese celebrate two birthdays; one is Yan Yat, considered the universal human birthday—it's on the seventh day of the Chinese New Year. On the other end of life, a special ceremony is held on the forty-ninth day after death—and forty-nine is seven times seven. The Japanese celebrate the seventh day after a baby's birth, and mourn on the seventh day and seventh week after a death.

Japanese folklore counts seven gods of luck. They're sometimes shown together on a treasure ship or they can be shown alone. They have a variety of magic implements, like a hat of invisibility, an inexhaustible purse, keys to the divine treasure house, a robe of feathers, and a lucky rain hat. They are the personifications of earthly happiness and good fortune; they include the god of wealth and farmers, the god of war and warriors, the god of fishermen, two gods of longevity, the god of happiness, and the goddess of music. They're related to the seven Buddhist Devas, who watch over human happiness and welfare.

## THE BLESSINGS OF SEVEN

For the Pythagoreans, seven was was an excellent number because it was the sum of three, the triangle, and four, the square—both perfect figures.

Seven isn't mathematically perfect as its predecessor, six, is, but seven has been called the number of perfection because God blessed the seventh day. Seven is the number of a variety of biblical feasts, and of sacrifice, purification, consecration, forgiveness, punishment, and reward in many religions.

Kwanzaa is not a biblical feast, but it is almost religious in nature. Over seven days, it celebrates the heritage of African Americans. First observed in 1966, Kwanzaa is partly a traditional harvest festival, but its emphasis is on the role of family and community in African American culture. It begins on December 26, and each of the next days is devoted to one of seven principles: unity, self-determination, collective work and responsibility, cooperative economics, purpose, creativity, and faith. Each day, one candle of a seven-branched candelabrum is lit. Gifts are exchanged, and a Karamu, or feast, is part of the festivities.

> There are seven sins in the world: Wealth without work, Pleasure without conscience, Knowledge without humanity, Commerce without morality, Science without humanity, Worship without sacrifice, and Politics without principle.
>
> MAHATMA GANDHI

For Jews, the seven-branched menorah is an ancient symbol. There was an oil-burning menorah in the Temple of Jerusalem; before that, one was used in the Tabernacle, the portable sanctuary. According to the Torah, God showed the design for the menorah to Moses. There's a theory that the seven branches represent the seven heavenly bodies that ancients counted in the sky—the sun, moon, and five planets. A seven-branched menorah is now part of the Israeli coat of arms. (The nine-branched menorah used during Hanukkah is waiting for us when we reach nine.)

## THE MOON'S SEVEN

Part of seven's enormous power, in so many places and religions, must have derived originally from the mysteries of the moon. Each of the moon's phases lasts about seven days, and before there were clocks to tick away our days, we learned to keep track of time by counting the number of days in each of the phases of the moon, and then the days between full moons, and the black nights of the new moon. We gave the moon names as it waxed and waned through the year, and lived by its light, shining—or not—on the fields and the seas, giving a pale glow for the long nights of the harvest, making shadows in the dark.

> **He will deliver you from six troubles; in seven no harm shall touch you.**
>
> **JOB 5:19**

Our weeks have seven days, like the phases of the moon, but it hasn't always been this way. Strangely, both the ancient Egyptians and the French under the revolutionary calendar counted ten days to each week. The seven-day week probably originated, as did so much else, in Mesopotamia. It took its rhythm from the moon, and it celebrated the seven celestial bodies that the earliest astronomers thought revolved around the earth, around us, as important as we thought we were. The seventh day was consistently the Sabbath day, but there wasn't much agreement about which day was seventh. For Jews, it was the last day of the week: Saturday. For Christians, it was the first day, Sunday. And for Muslims, it was the sixth day, Friday.

The names of the days of the week come from a profusion of sources, mostly Latin, especially for the Romance languages, and

Nordic mythology, sometimes filtered through Germanic languages and Old English.

In Latin, Sunday was *Dies Solis,* the day of the sun, and in Old English that became *Sunnandaeg,* thus Sunday. Monday is the moon's day, from the Old English *Monandaeg*—in German, *Montag;* in Latin, *Dies Lunae,* the day of the moon; and in French, *Lundi.*

The Romans called Tuesday *Martis,* after Mars, the Roman god of war; in French, then, *Mardi,* and thus the New Orleans festival *Mardi Gras*—Fat Tuesday, the last day before the lean days of Lent. Tuesday links to the Germanic god Tiu, in charge of war and the sky, and also to the Norse god Tyr; in Old English, Tiw—thus *Tiw's daeg.*

For the Romans, Wednesday was *Dies Mercurii,* in tribute to Mercury; thus the French *Mercredi.* In German, Wednesday is *Mittwoch*—midweek, from the custom of calling days by their number. The English Wednesday nods to the Norse Odin (translated in German to Woden), the god of storms; the Old English was *Wodnesdaeg.*

Thor was the Norse god of thunder, and so we have *Thor's daeg.* The Romans gave the day to Jove, their god of thunder, and so it was called *Dies Iovis;* in French, *Jeudi.* Friday gives us a woman for the first time: Frigg was Odin's wife and Thor's mother; her day is *Frigg's daeg.* For the Romans, it was *Dies Veneris,* after Venus, the goddess of love; in French, *Vendredi.* For the Romans, Saturday was *Dies Saturni,* the day of Saturn, the Roman god of agriculture. The Old English *Saeternesdaeg* derives from Saturn too, and, in order, so does Saturday.

Monday's child is fair of face;
Tuesday's child is full of grace.
Wednesday's child is full of woe;
Thursday's child has far to go.
Friday's child is loving and giving;
Saturday's child works hard for a living.
But the child that's born on the Sabbath day
Is bonny and blithe and good and gay.

OLD RHYME

And then, it's Sunday again; seven days have gone by, and the week begins again.

Our lives are spent in the rhythm of those seven days, under the moon and its phases, the sun, and the planets. No wonder seven was so long considered lucky—even magical. To be a seventh child is high good fortune, and to be a seventh son is almost as good as it gets. In a world ruled by men, in which Wednesday and Thursday—the days of the husband and of the son—come *before* the day of the wife and the mother, in which sons grew up to work and to inherit and daughters to be decorated and married away, best of all in that world, best of all that could be imagined and wondered at, was to be the seventh son of a seventh son, a healer, a magician, with powers granted by the gods. A multiple of sons, a multiplication of sevens, a deep magic for the boy child.

**"Seven years and six months!" Humpty Dumpty repeated thoughtfully. "An uncomfortable sort of age. Now if you'd asked my advice, I'd have said, 'Leave off at seven.'"**

**LEWIS CARROLL,** *THROUGH THE LOOKING-GLASS*

# 8

## EIGHT IS FOREVER

Eight is a quirky number. I don't mean that it's mixed up—it's not; it's solid and respectable and it knows where it's going (to nine, undoubtedly); it's just quirky is all. In the case of eight, quirky is a nice thing to be (as it so often is). After all, eight is twice four, and four is twice two, and two is twice one. That's a proud heritage.

> Man is not an end but a beginning. We are at the beginning of the second week. We are children of the eighth day.
>
> THORNTON WILDER

Eight is also a deliciously round number. Like zero, it is nothing if not round. The numeral 8 stands like a snowman on a field of white paper, loop on top of loop. True, it has no carrot nose or raisin eyes; instead there are two delicious, pattable tummies, round and chubby, soft, yet firm. For Freud, perhaps, two delicious, pattable breasts—or the balls from which the world hangs. Indeed, if you turn that zaftig numeral on its side, it becomes the mathematical symbol for infinity: $\infty$. Infinity could be a wonderfully Freudian idea—think of a baby, lost in the timelessness of its mother's breasts, milky, warm, sweet, safe, forever.

That sleeping eight is no more or less than two joined circles, looping from one to the other, making their magic crossing in the middle without stopping, without lifting pen from paper, without beginning or end, simply traveling, around and around, forever—and that *is* infinity.

There were times when the symbol for infinity stood erect—8—rather than on its side. The sideways eight (called the lemniscate) appeared formally for the first time in a list of signs compiled by an English mathematician, John Wallis, in 1655. It has been with us, one way or another, for a very long time, and it has had only a few variations along the way. It was used in astronomy, magic, and mysticism in many different civilizations. Sometimes it was simpler, a charm that stood for eternal union or infinite happiness. Often it was an S, or it looked like a serpent. For the Celts, it was an S that represented a serpent which bites its own tail. For the Assyrians, too, it was "the serpent of eternal life." In *The Universal History of Numbers,* Georges Ifrah writes that the serpent sign has been "connected to ideas such as the sky, the universe, the axis of the world, the night of beginnings, the primordial substance, the vital principle, life, eternal life, sexual energy, spiritual energy, vestiges of the past, the seed of things to come, cyclical development and resorption, longevity, extreme fertility, the incalculable quantity, abundance, immensity, totality, absolute stability, endless movement," and, he writes, "etc."

> **Infinity, the inexhaustibility of the counting process, is a mathematical assumption, the basic assumption of arithmetic, on which all mathematics rests.**
>
> **TOBIAS DANTZIG,** *NUMBER: THE LANGUAGE OF SCIENCE*

Numbers are the language of infinity, assuming numbers can be used to count themselves. There's no place where the last number can be reached, no end to the string of one more and one more and one more. Whatever number you're at is one less than the next number. Add one more, and you have another number, larger than the last. Numbers *are* infinity—numbers without end.

In India, where scholars say our numerals began, infinity is poetically defined in part as "the sum of all the drops of rain

> Man is equally incapable of seeing the nothingness from which he emerges and the infinity in which he is engulfed.
>
> BLAISE PASCAL, *PENSEES*

which, in ten thousand years, would fall each day on all the worlds." Raindrops, grains of sand, or stars often symbolize infinity—though the concept of infinity goes beyond even those; it is unimaginably vast, a number, if it *is* a number, that can never be reached. For some, it was a number only for God to know, and it represented God as infinitely powerful.

The word *infinity* comes from the Latin *infinitus,* "that which has no end" or "that which never ends." Grains of sand can be counted—given enough time, endless patience, and a sturdy sieve—but infinity is counting without end, one added to the last number, forever. In that way, one, powerful, potent one, is all. But it isn't one that stands opposite to infinity. The opposite of infinity is zero, nothing at all.

## SON OF BONACCIO

Eight is the last of the single-digit Fibonacci numbers; 1 and 1 again, 2, 3, 5, and 8 are the first in the series. If you note immediately that each number is the sum of the two numbers that came before it (1+1=2; 1+2=3; 2+3=5; 3+5=8), you will have grasped what a Fibonacci number is—one in a chain of numbers, each based on the sum of the two previous numbers, and you'll know that the next one is 13, because 8+5=13. Why Fibonacci? And so what? Ah, there lies a tale.

The story begins with a medieval customs official, Guglielmo Bonaccio, who was secretary to the Republic of Pisa. When his son, Leonardo, was born in about 1175, Guglielmo wanted him to grow up to be a merchant, so when Guglielmo became responsible for the Pisan trading colony in Algeria, he brought Leonardo along and arranged for him to be tutored in the art of calculation, to help him be a better businessman. "He made me learn how to use the abacus when I was still a child," Leonardo later wrote, "because he saw how I would benefit from this in later life. In this way I learned the art of counting using the nine Indian figures. . . . That is why, with these nine figures, and with the sign 0, called *zephirum* in Arabic, all the numbers you may wish can be written."

Because his teachers were Muslims, Leonardo learned the Hindu-Arabic numerals, still unknown in Europe, where Roman numerals were the numbers of the day. He learned about the place value of numbers—that, reading from right to left, one can read numbers as units, one through nine, then as tens, and then as hundreds, so that simply by glancing at a number like 323, one is able to understand easily and quickly that the number is three hundred and twenty-three.

Traveling around the Mediterranean, he learned algebra, compared several numerical systems, and learned to use the abacus as a much better way of working with numbers.

In about 1200, Leonardo returned to Italy. Today we would call him a nerd, with a row of pens and pencils in his shirt pocket. Then, Pisans called him *Bigollone*—a sort of blockhead, a man of no importance. He was supposedly often seen walking around the city, stopping frequently to stand lost in thought for a moment and then scribbling Arabic numerals on the nearest wall with the pieces of chalk that were always in one pocket or another.

History has given Leonardo a slightly different name—not *Bigollone,* but *Fibonacci,* in Latin, *filius Bonacci,* son of the Bonaccio family. Today his name reverberates through mathematical history, from Pisa to the whole world.

In 1202, when he was twenty-seven, Leonardo Bonaccio Fibonacci published his first book, *Liber abaci* (the Book of the Abacus or the Book of Calculations). In it, he explained how to

> If by chance I have omitted anything more or less proper or necessary, I beg forgiveness, since there is no one who is without fault and circumspect in all matters.
>
> **FIBONACCI,** *LIBER ABACI*

translate from Roman numerals to the new Hindu-Arabic numerals, how to use them, and the ways in which they made arithmetic possible. He used as examples a range of problems, many having to do with business practices of the day. And he phrased his examples with words rather than with numbers, so that his algebra was more easily understood.

In *The Golden Ratio,* contemporary writer Mario Livio quotes an example of one of Fibonacci's problems:

> A man whose end was approaching summoned his sons and said: "Divide my money as I shall prescribe." To his eldest son, he said, "You are to have 1 bezant [a gold coin first struck at Byzantium] and a seventh of what is left." To his second son he said, "Take 2 bezants and a seventh of what remains." To the third son, "You are to take 3 bezants and a seventh of what is left." Thus he gave each son 1 bezant more than the previous son and a seventh of what remained, and to the last son all that was left. After following their father's instructions with care, the sons found that they had shared their inheritance equally. How many sons were there, and how large was the estate?

(If you need to know and would rather not get involved with *x,* the answer is that there were six sons and the estate comprised thirty-six bezants.)

*Liber abaci* was Fibonacci's first book, and one of four to survive. Before Gutenberg, books had to be hand-copied, and only a few copies would be made. Fibonacci's other books are on commercial arithmetic, on geometry, and a commentary on Euclid's works, but *Liber abaci* is the book that made him famous.

## RABBITS AND MORE RABBITS

*Liber abaci* did more than bring Arabic numerals to Europe, which is important enough. It also posed a theoretical problem, and offered its solution.

Suppose, Fibonacci wrote, someone placed a pair of rabbits in an enclosed space, and left them to breed for one year, presumably taking good care of them in the interim. ("A certain man put a pair of rabbits in a place surrounded on all sides by a wall. . . .") If every month, a pair of rabbits produced another pair, and each pair—including the original two—had offspring beginning two months after their own birth, how many pairs of rabbits would there be at the end of a year? (We'll assume also that each pair meant one male and one female, and that the rabbits reproduced with or without incest, even as did Adam and Eve and their surviving sons.)

> **All the thoughts of a turtle are turtles, and of a rabbit, rabbits.**
>
> RALPH WALDO EMERSON, "THE NATURAL HISTORY OF INTELLECT"

The answer is that there would be 233 pairs of rabbits at the end of a year. But that's not the important part of what Fibonacci did. He noted that in the first month, there would be no new rabbits, just the original one pair, but at the end of the second month, there would be a new pair, and another new pair at the end of the third month. By the fourth month, the first offspring would begin having their own baby rabbits, and every month after that, there would be new rabbits arriving in ever-increasing numbers. This is what the sequence of rabbit pairs would look like, in numbers: 0, 1, 1, 2, 3, 5, 8, 13, 21, 34, 55, 89, 144, 233, 377 . . . and so on, assuming the rabbits lived forever.

That sequence (formally called the *Fibonacci sequence* for the first time by a French mathematician in the nineteenth century) is noteworthy in two ways:

First, each number is the sum of the preceding two numbers: 2=1+1; 3=1+2; 5=2+3; 8=3+5. . . . No matter how high the numbers go, each number is still the sum of the two previous numbers.

And then—and then—there's the number that is almost magic—no, it *is* magic; it applies to a vast range of natural and man-made shapes. It's arrived at by dividing any number in the Fibonacci sequence (after the tenth, because before that the numbers aren't quite large enough) by the number just before it. The answer comes closer and closer to 1.6180339 as the numbers go higher, and the ratio of 1.6180339 to 1 turns out to be the mathematical basis for everything from the shell of a snail to the tails of spiraling galaxies. It's called the *Golden Ratio* or the *Golden Mean,* a remarkable, universal, and beautiful number, one of the numbers, said the *New York Times,* "that seem to be encoded within the software of the universe."

> I had a feeling once about Mathematics—that I saw it all. Depth beyond Depth was revealed to me—the Byss and the Abyss. I saw—as one might see the transit of Venus . . . a quantity passing through infinity . . . but it was after dinner and I let it go.
>
> WINSTON CHURCHILL

The ancient Egyptians knew about the Golden Ratio without knowing exactly what it was. They based the shape of the pyramids on a rough proportion of 5 to 8 (two numbers in the Fibonacci sequence that are close to the Golden Ratio). The average ratio of the height of the Pyramids at Giza to their base is about 5 to 8. There are scholars who believe that the original height of the Great Pyramid would be translated as 484 feet and 5 inches (the original height has to be estimated because of the effects of age on the stones), or 5,813 inches (5, 8, and 13 are Fibonacci numbers). If the height of the Great Pyramid is thought of as the radius of a circle—a straight line extending up from the base—the circumference of that circle would be 36,524.2 inches. The precise length of a year is 365.242 days.

The Greeks—again hundreds of years before Fibonacci and without his numbers—knew that the Golden Ratio determined which shapes were simply the most pleasing to the eye; they arrived at the proportions through geometry rather than arithmetic, and they called the result the *Divine Section.* (They were right about the shapes that please the eye. In a recent study, hundreds of people looked at rectangles of various proportions and shapes. Seventy-five percent preferred rectangles whose size was based on the Golden Ratio. The Divine Section is found, to this day, in the dimensions of thousands of objects we use all the time—windows, playing cards, writing pads, and books among them.)

When a line is divided into two segments, and the entire line is to the larger segment as the larger segment is to the smaller, this is the Golden Ratio, 1.6180339 to 1. Euclid described the calculation in his *Elements;* the front of the Parthenon, including the original pediment, was based on Golden rectangles. Greek sculptors used the Golden Ratio in divisions of the bodies they created. Greek vases include precise Golden proportions.

Leonardo da Vinci used the Golden Ratio in his painting and sculpture, and collaborated with a monk to produce a book titled *De Divina Proportione.* Many of art's masters used the Golden Ratio again and again, from Piero della Francesca and Bellini to Poussin and, in the nineteenth century, Seurat.

In the seventeenth century, Jakob Bernoulli made the connection between the Golden Ratio and nature itself. He did this by means of a rectangle, divided again and again into its Golden Ratio measurements. If a rectangle is divided by the Golden Ratio, and a square is drawn into the smaller section, another, smaller, Golden rectangle will be left. Again, if that smaller rectangle is divided by the Golden Ratio and another square drawn inside it, there will be a third, smaller rectangular space left over—and so on, until the space is too small to comfortably divide with a pen or pencil. If the center points of all these squares are connected by straight lines, the lines form a sort of stiff spiral shape. And if the spiral is shaped into a curve, rather than a series of lines, the result is what's called a *Golden Spiral.* (Bernoulli liked this shape, and some of its attributes, so much that he ordered it engraved on his tombstone.)

Bernoulli's Golden Spiral existed long before he discovered it. Like America before Columbus, it was always there. The chambered nautilus builds its shell—and also its shell's inner partitions—in the shape of the Golden Spiral. In much more primitive forms, tiny plankton build their bodies in that same spiral, as does the garden snail. The shore of Cape Cod is a Golden Spiral; a breaking wave—under which surfers ride, oblivious to geometry—is a Golden Spiral. Sunflower seeds are arranged in Golden Spirals, clockwise and counterclockwise. More: most sunflower heads have spirals made up of thirty-four and fifty-five seeds; small ones have spirals of twenty-one and thirty-four seeds; larger ones have fifty-five and eighty-nine. All Fibonacci numbers.

The eyes of pineapples are in the middle of scaly plates that whirl around the fruit in spirals of five in one direction, eight in another, and thirteen in a third. Pine needles grow in clusters of two, three, or five, depending on the species, and their cones are Fibonacci spirals as well.

Starting at the bottom of the stem of almost any green plant, and counting the leaves going up the stalk until a leaf directly above the bottom leaf is reached, without including the first leaf, will give a Fibonacci number. And the number of times the stem is circled on the way up—either clockwise or counterclockwise—yields another Fibonacci number.

The seeds of raspberries, the petals of most flowers . . . all Fibonacci numbers. There are a few flowers with one petal (calla lilies, for example) or with two, more with three, but four are rare. Find a four-leaf clover for luck—there aren't very many. Hundreds of flowers, wild and cultivated, have five petals. Black-eyed susans have thirteen petals; columbine has five; shasta daisies have twenty-one. . . . Ordinary field daisies have thirty-four petals. (So if you start with "she loves me," you'll end with "she loves me not"—except that there are exceptions. A few have either thirty-three or thirty-five, a Fibonacci error and a plus for love.)

A ram's horn, a lion's claw, a parrot's beak, an elephant tusk—all are Golden Spirals. Some galaxies send their arms outward in vast clusters of stars, based on the same Golden Spiral, the same Fibonacci numbers. From the common garden snail to the elephant's tusk, from our own Milky Way to the farthest reaches of the universe, we are all tied together by Fibonacci numbers and the Golden Ratio. The Golden Spiral is nature's way of preventing overcrowding, of keeping order and maintaining economy. It is also a perfectly beautiful line.

Less grandly, Fibonacci numbers provide an early clue in Dan Brown's blockbuster novel, *The Da Vinci Code.* And they've gone on to inspire a rash of poems—called "Fibs"—based on the Fibonacci pattern. They usually have six lines, with the number of syllables in each line equaling the sum of the syllables in the two previous lines. The first two lines have one syllable each; the third line has two syllables; the fourth, three; the fifth, five; and the sixth, eight. For example:

*Now*
*I*
*look for*

*the numbers*
*of Fibonacci*
*instead I find the universe*

## AND BACK TO EIGHT

Eight didn't start out as one circle on top of another; rather, the first numeral eight, in India, looked like a very modern chair—like an upper-case H, with the bottom half of the first vertical stroke and the top half of the second both missing, and the rest, the chair, done in an easy curve. This pleasing line evolved until it looked more like an S, and there was a danger of confusing it with the numeral 5. The Arabs resolved the confusion by connecting the beginning and the end of the S, and Europeans finished the process by rounding the top and the bottom.

> To speak algebraically, Mr. Matthews is execrable but Mr. Channing is (x + 1) deplorable.
>
> EDGAR ALLEN POE, DESCRIBING TWO WRITERS

The names for eight were remarkably consistent for a long time: the Indo-European prototype was *okto,* and so was the Greek. In Latin, eight became *octo.* Dutch eight is *acht;* French, *huit,* Irish, *ocht,* Swedish, *atta,* and German, like Dutch, *acht.* The English eight traveled sideways just a bit and came from the Middle English *eighte* and the Old English *eahta.*

The Latin *octo* gives us a small army of words beginning with those two syllables. If you're eighty or over, you're an octogenarian. An octopus has eight tentacles, and an octet is eight musicians—or it's the music that's written for them.

An octothorpe is a fancy name for the pound or number symbol—#. According to *A Word a Day,* in the early 1960s Bell Labs wanted a new name for the # symbol in order to use it on their new touch-tone keypads. (They needed two new keys; the other is the asterisk, which they called the star.) There are several stories about where the name *octothorpe* came from. Everybody agrees about the *octo* part; there are eight tips to the # symbol. It's the origin of *thorpe* that varies, but the best (and the only documented) story has it that the engineer who was working on the new phones belonged to a group seeking the return of Olympic athlete Jim Thorpe's medals. (Thorpe had been disqualified on the basis of his professional status; his medals were restored posthumously.) So the engineer put *thorpe* on the end of *octo.* I'd say the name stuck, but does anyone ever use it? (The pound sign, according to *A Word a Day,* is also known as a crosshatch, hash, crunch, mesh, hex, tic-tac-toe, flash, grid, gridlet, pigpen, gate, hak, oof, rake, fence, square, and widget mark, and in music it's a sharp sign.)

Octagons are eight-sided figures, like stop signs on street corners everywhere. One of the nice things about eight is that an octagon is the first form to begin changing a square into a circle—it's almost round.

October, I hear you saying. What about October? Hardly the eighth month. Well, yes, but once it was, even though it's now the tenth—more about this and all the other mixed-up months when we get to twelve, the number of months, no matter what their names may be, in a year.

## MUSIC, MUSIC, MUSIC

The most common musical scale—a word derived from the Italian *scala,* "ladder"—has only five notes (C, D, F, G, and A). This pentatonic scale is used in the music of many ancient cultures, from Chinese and Africans to Celts, Scots, and American Indians. The scale now used in Western music is the diatonic scale. (The black notes on a piano—which has eighty-eight keys—correspond to a pentatonic scale; the white notes, diatonic.) In the diatonic scale,

> Music is given to us with the sole purpose of establishing an order in things, including, and particularly, the coordination between man and time.
>
> IGOR STRAVINSKY

our do-re-mi, an octave is eight tones above or below any given note. Actually, I blush to admit, there are only seven notes in an octave, but the beginning note and the final note are eight tones apart—in effect, repetitions at another pitch. As: do, re, mi, fa, sol, la, ti, and—once again—do. The last note, the eighth tone, also becomes the first note of the next octave.

That scale, the do-re-mi, wasn't always called do, re, mi—*The Sound of Music* notwithstanding. *Ut* was once the name for the lowest note, and *si* was the highest—so the scale was ut, re, mi, fa, sol, la, si, ut, all derived from the initial syllables of a Latin hymn. *Ut* and *si* changed into *do* and *ti,* but *ut* left its mark: the word *gamut,* which means the complete range of something, is from medieval Latin; it's a contraction of *gamma*—the third letter of the Greek alphabet, which was used to represent the lowest tone—and *ut,* the lowest note in an octave. *Gamma* + *ut,* on the one hand, contracted to form *gamut,* and on the other, expanded to mean all the notes—the complete range from *ut* to *ut,* or from *do* to *do,* from top to bottom and everything between.

Music is full of complications, and many of them are mathematical. An easy example: the ratio of frequencies of two notes an octave apart is two to one because the interval between the notes is either double or half of the original frequency. If one note has a frequency of 400 hertz (a measure of pitch), the note an octave above it is at 800 Hz, and the note an octave below is at 200 Hz. That's just the beginning, the barest surface. Music theory is full of mathematical relationships, in note lengths, pitch, rhythm, harmony, time signatures, volume, timbre, and counterpoint. The intervals between notes, consonance and dissonance when different notes are played together, are all mathematically determined. There is frequently a strong connection between mathematicians and musicians—the love of one field often relates to the love of the other. From the outside, it's possible to see that there is a kind of music—from Pythagoras on—in math, and a kind of mathematics in music. A Bach fugue is made up of magic—but it can be graphed mathematically even while it is pure music, brilliant and uncanny in its perfection.

The most overt connection between music and math can be found in Joseph Schillinger's "System," the application of mathematical and scientific logic to music. Schillinger, who taught at Columbia University, believed music was pure math—he claimed he could write a musical composition based on the morning stock exchange report. He believed in the universality of mathematical patterns—they applied to the structure of our nervous system, to the patterns in linoleum, and, almost above all, to music.

George Gershwin was one of Schillinger's star pupils, and wrote *Porgy and Bess* while studying with him. Benny Goodman, Oscar Levant, Tommy Dorsey—all Schillinger students. Another, Lawrence Berk, founded a music school in Boston in 1945 to teach the System to others; the school became the Berklee College of Music, and the Schillinger System was taught there until the 1960s. Another pupil, Glenn Miller, once wrote a series of harmonic exercises as homework. Schillinger studied the notes and suggested that Miller orchestrate what he had written. The result was "Moonlight Serenade," Miller's theme song.

## BELIEVING IN EIGHT

We see more and more in the crowded heavens beyond our earth as our vision becomes clearer and stronger. But once, long before astronomers were able to see beyond Uranus, some believed that

> Fall seven times; stand up eight.
>
> **JAPANESE PROVERB**

beyond the seventh planet, there was an unseeable place, an *eighth* place, a sphere where the stars were, and where the gateway to paradise could be found.

The association of eight with paradise exists in many religions. In Islam, there are seven hells and eight paradises, indicating that God's mercy is greater than his wrath. Many Muslim gardens are divided into four or eight parts, symbolizing the beatitude of heaven. Along with the eight paradises, Islamic mythology also has it that eight angels carry the divine throne.

Buddha taught the Noble Eightfold Path as the way of suppressing the suffering which is all life, according to the Four Noble Truths. The path is sometimes called the Middle Path because it lies between the sensual pleasures of the body, materialism, and the self-mortification of asceticism.

There are eight precious items of Confucianism, and sixty-four (eight times eight) configurations in the *I Ching,* the Book of Changes, the ancient and classic Chinese text. Legend tells us that the form and principles of the *I Ching* were developed by the Emperor Fu Hsi, who ruled in the third millennium BC—five thousand years ago. According to one legend, Fu Hsi based the hexagrams on a design he saw on the back of a tortoise near the banks of the Yellow River. Confucius is supposed to have written notes on the *I Ching,* and the book is one of the Five Confucian Classics. It centers on the ideas of the balances of opposites (like yin and yang), change as an inevitable process, and the importance of the acceptance of change as well as of persistent principles that never vary. All of this is basic to Chinese culture and beliefs. Around the world, many people believe that the *I Ching,* with its sixty-four symbolic hexagrams, offers a way of seeing the future, even of affecting it. It certainly offers a means of clarifying one's thoughts about the present.

The flag of South Korea is the tai chi symbol (*taeguk* in Korean), the circle of yin and yang in perpetual and dynamic balance, surrounded by four of the trigrams from the *I Ching,* for heaven, water, earth, and fire.

Christians believe that, according to the Gospel of Saint Matthew, Jesus uttered eight blessings at the beginning of the Sermon on the Mount—the Beatitudes—though the Gospel of Saint Luke lists four blessings and four woes. The resurrection of Christ was on the eighth day of the Passion. Some say the seven days of the week are the time of this world, but the eighth day is the time of everlasting life.

In Jewish tradition, the eighth day after birth is the day for the circumcision of male babies, the physical recognition of the covenant God made with Abraham. Another eight is Hanukkah, the Feast of Lights, which lasts eight days. Each evening at sunset, the candles of the eight-branched candelabrum, the menorah, are lit in sequence: one candle on the first night, two on the second, three on the third, until on the last night, all eight are aglow. (A ninth candle is set apart from the others and is used to light them.) The candles commemorate the miracle of the oil—the miracle of the Temple lamp. The lamp contained only enough holy oil for one day, but it burned for eight days, until a supply of new oil could be obtained. Hanukkah celebrates the rededication of the Second Temple of Jerusalem in 164 BC, after the Temple's desecration by the Syrians.

Blessed are the poor in spirit: for theirs is the kingdom of heaven. Blessed are the meek: for they shall possess the land. Blessed are they who mourn: for they shall be comforted. Blessed are they that hunger and thirst after justice: for they shall have their fill. Blessed are the merciful: for they shall obtain mercy. Blessed are the clean of heart: for they shall see God. Blessed are the peacemakers: for they shall be called the children of God. Blessed are they that suffer persecution for justice's sake, for theirs is the kingdom of heaven.

THE BEATITUDES

## ASSORTED EIGHTS

There's another side to eight, of course. There's always another side. To be behind the eight ball, for instance, is not a good place to be. The expression comes from the eight ball game of billiards. The player who lands the (black) eight ball in a pocket too early, or who misplays it, loses the game.

In another game, we have the Black Sox scandal of 1919, when eight players on the Chicago White Sox—that year's American League champions—conspired to throw the World Series to the Cincinnati Reds. Best known of the eight was Shoeless Joe Jackson ("Say it ain't so, Joe!"), but all eight were forever banned from baseball in 1921.

A Magic Eight Ball is another story. It answers questions—magically, of course. It's a hollow plastic ball filled with blue liquid, with a window on one side. Inside floats a white plastic shape with answers to almost any possible yes-or-no question on its various sides. You hold the sphere with its window down and ask your question (Does she love me?). When you turn the ball over, the shape floats to the top and presses one of its sides against the window. Ten of the possible answers are variations on yes (including Yes—Definitely); five are ambiguous (Better Not Tell You Now); and five, alas, are variations of no (Outlook Not So Good).

In bingo slang, before we learned not to say such things, eight was called "one fat lady," and eighty-eight was two of the same.

> Reasons for Preferring an Elderly Mistress: Eighth and lastly: They are so grateful.
>
> BENJAMIN FRANKLIN

There are indeed eight vegetables in V-8 juice. Spiders—like most arachnids—have eight legs. In Chinese, eight is considered a lucky number because the word for eight sounds like the word for wealth. The Beijing Summer Olympics are scheduled to open at 8 p.m. on August 8—8/8/08.

If you remember *M*A*S*H,* you

probably remember what a Section Eight is—a discharge from the military because of being mentally unfit for service. Thus, Jamie Farr and his lovely dresses. (A catch-22, from the book of that name: if you *want* to be discharged, you can't be mentally unfit.) Section Eights have been replaced by a variety of other regulations.

Happier eights are "pieces of eight"—from the silver coins, *pesos* (literally, "weights"), minted after 1497 by the Spanish, after they had found extensive silver deposits in Mexico and South America. Often the coins were cut into four or eight pieces in order to make change—they were pure silver, so the pieces were worth their weight, just as the whole was. Pieces of eight became the whole coin; two bits (or two pieces—or, formally, two *reales*) became what we now call a quarter. The peso was roughly equivalent to the silver *thaler* minted in Bohemia beginning in 1517; the name for the Bohemian coin was *thaler; thaler* became *dahler* in Low German, and eventually dollar in English.

Because the American colonies had a shortage of English currency before the Revolution, the colonists began using the Spanish coins, calling them dollars. Prices on the New York Stock Exchange were noted in one-eighth-of-a-dollar denominations, matching the pieces of eight that a peso or dollar could be broken into; eventually, the stock exchange converted to sixteenths of a dollar, and finally to decimal figures and the tenths of a dollar that are used now.

We work for eight hours a day, but the eight-hour workday wasn't always as nearly sacrosanct as it is today. There was a time when workers struck to get a ten-hour workday. In 1886, the Haymarket Square Riot focused on demands for shorter workdays; several people in the crowd and seven policemen were killed in the Chicago uprising. The first federal law to establish an eight-hour day passed in 1916; it was for railroad workers. The Fair Labor Standards Act passed in 1938.

Writers—some of whom work eight-hour days—may know

(but probably don't) that there are eight parts of speech in traditional standard grammar: nouns, pronouns, verbs, adjectives, adverbs, prepositions, conjunctions, and interjections. What are interjections? Exclamatory words like Oh! as in Oh! I don't know!

A few more eights before we finish: Hawaii, which became a state in 1959, contains eight major islands: Hawaii, Oahu, Kahoolawe, Kauai, Lanai, Maui, Molokai, and Niihau. There are eight Ivy League schools: Harvard, Yale, Pennsylvania, Princeton, Columbia, Brown, Dartmouth, and Cornell—listed in chronological order, as Harvard was founded in 1636, and Cornell in 1865.

In the same way as an octave is made up of seven notes plus the first note repeated (remember that eight is a quirky number), a week in Germany is called *acht tage,* "eight days," and in France, *huit jours,* "eight days" again. A week in both countries still has seven days, but it takes eight days to arrive back at the same day a week later: Sunday, Monday, Tuesday, Wednesday, Thursday, Friday, Saturday, and Sunday again. Eight days: a week.

# 9

## NINE IS A NEW NUMBER

You'd think because we counted with our fingers for so long that the progression of numbers from one to ten would be natural and easy—even inevitable. Not so. Thinking about numbers has never been easy, even before anybody knew anything about algebra and calculus. Simply learning to count, one by one, took a very long time, and we approached it in a great many different ways.

> Wandering in many a coral grove, Fair Nine, forsaking Poetry!
>
> **WILLIAM BLAKE, "TO THE MUSES"**

Going from one to two was fairly easy. The next leap we made was to three, the number which is one beyond a pair. Counting to four seems like a natural next step, but it's natural only given the luxury of looking backward.

Having reached that far, we stopped for a while, with the convenient *many* for everything that came after four. After that, in a variety of cultures, the count went to eight, using both hands but ignoring the thumbs, as if all they were good for was sucking. But inevitably there was a need to count beyond eight. When the next number was given a name, at different times and in different places, it often meant something like "the new number."

Words stand as a testament to this sense of nine's newness as a number, after the fingers—not counting the thumbs as fingers—had been used up. In Sanskrit, "nine" is *nava,* while *navas* is "new." In Latin, *novem* is "nine" and "new" is *novus.* A nova is a new star; a novice is new to something; a novel took its name from the idea of a new story; and a novelty is something new and unusual. To innovate is to create something new; to renovate is to make new again. The same link exists in Gothic; *niun* is "nine" and *niujis* is "new." The Indo-European prototype for nine is *newn.* Even in Egyptian, nine is a cousin of new: the Egyptian word for nine means both the rising of the sun in the east and the first appearance of the new moon.

## NINE TIMES NINE

At heart, nine is really three threes. That makes it another mystical and potent number—and it has plenty of magic in its own right as well. Like 6, it's a perfect little sperm, but now swimming upstream toward the egg that is zero. And it's full of magical mathematical tricks, singing its chants—the times table—as it goes. This is what it says:

$$
\begin{aligned}
0\times9&=\ 0\\
1\times9&=\ 9\\
2\times9&=18\\
3\times9&=27\\
4\times9&=36\\
5\times9&=45\\
6\times9&=54\\
7\times9&=63\\
8\times9&=72\\
9\times9&=81\\
10\times9&=90
\end{aligned}
$$

There are four ways of looking at that column of numbers. Left to right: the first column (the number that nine is being multiplied by) goes from zero to ten, in order, from top to bottom. Okay, that's the way it usually is. But moving right along, the first column to the right of the equal sign (not counting the first two lines) also goes in order from one to nine. The final column, on the right, goes backward from zero to nine to zero again. Finally, if you add the digits in the answers, the sum is always nine. (Two plus seven equals nine; three plus six equals nine; four plus five equals nine.) *And* one extra—ta da—the first digit to the right of the equal sign is always one less than the number you're multiplying by. (Two times nine equals eighteen, or one-eight. Three times nine equals twenty-seven, or two-seven.)

> [Nine is] so much more than poor old Eight; a darling little number in its own right of course, and a pleasant sort of number, but not Nine. You'd go to bed with Eight, but you'd marry Nine. The sheer, cunning, marvelously bumsucking, miraculously bloody nineness of it.
>
> **JOSEPH O'CONNOR,** "THE SCHOOLMASTER"

Rabbi ben Ezra (of whom Robert Browning wrote) devised a diagram to show nine's magic tricks. It shows the same multiplication table in a different form.

| | | |
|---|---|---|
| | $9=9\times 1$ | |
| $9\times 9=81$ | $8\leftrightarrow 1$ | $18=9\times 2$ |
| $9\times 8=72$ | $7\leftrightarrow 2$ | $27=9\times 3$ |
| $9\times 7=63$ | $6\leftrightarrow 3$ | $36=9\times 4$ |
| $9\times 6=54$ | $5\leftrightarrow 4$ | $45=9\times 5$ |

His neat diagram sends nine up and down and left and right and all around. Next door to magic.

The mystical Gematria (discussed in the chapter about six) also had something to say about nine. In Hebrew, the numerical values of the word for truth add up to 441, and $4+4+1=9$. Thus, for Kabbalists following Gematria the number nine shows that God is truth, and shows further the invariance of God, for $9\times1=9$ and $0+9=9$; $9\times2=18$ and $1+8=9$; $9\times3=27$, and $2+7=9$, and so on, as we've already seen, through the nine times table to $9\times9=81$, and $8+1=9$. Nine is the constant and God is constant.

## MAGIC

All of this gives nine a lot to sing about. Indeed, it could sing about much more—being one less than ten gives it all sorts of abilities other numbers don't have. Perhaps the most amazing are magic squares. Each magic square is made of nine boxes, like a framed tic-tac-toe diagram. A different number from one to nine is placed in each box, and no matter how you add the boxes—across, down, diagonally—the answer is always the same. Some magic squares have more than nine boxes and use more than nine numbers, but the first magic squares were always based on nine. If five goes in the middle with even numbers in the corners and the remaining odd numbers arranged in the remaining boxes, the answer to the totals of the rows, when the numbers are correctly placed, is always fifteen.

Sudoku puzzles are related to magic squares and are also similar to a puzzle created in the 1800s called Latin squares. Each puzzle has a nine-by-nine grid of squares, within which are nine three-by-three squares. The puzzles come with numbers already in some of the squares; the object is to fill in all of the squares so that every horizontal and vertical line—as well as every three-by-three square—contains the numbers one through nine with no number repeated.

> Ninety-nine percent of the world's lovers are not with their first choice. That's what makes the jukebox play.
>
> WILLIE NELSON

The first puzzles appeared in 1979 in an American puzzle magazine; they were called "Number Place." By 1984, they were appearing in Japan, and they spread around the world from there. No language is needed to work on sudoku—just logic and patience. In Japan, the puzzles are sometimes still called Number Place; everywhere else, the Japanese word—*sudoku* means "single number"—is used.

A London newspaper reports that there has been a 700 percent increase in pencil sales because of the sudoku craze. The BBC notes that a puzzle editor for one London newspaper published a puzzle with more than one solution, and received sixty thousand e-mails about it. In 2005, according to the *New York Times,* British Airways sent a notice to its thirteen thousand flight attendants, forbidding them to work on sudoku puzzles during takeoffs and landings—they're too distracting. In the United States, at the beginning of 2007, Amazon listed six hundred and fifty books about sudoku; total sales so far are well into the millions. According to *USA Today,* in the summer of 2005 seven of the top hundred best-selling books were compilations of sudoku puzzles. Worldwide, there are now daily puzzles in vast numbers of newspapers. Two mathematicians (Bertram Felgenhauer in the Department of Computer Science at the University of Dresden in Germany and Frazer Jarvis in the Department of Pure Mathematics at the University of Sheffield in the UK) have estimated that there is a huge number of possible sudoku grids—6,670,903,752,021,072,936,960 to be precise. Inevitably, there has been a World Sudoku Championship competition—the first was held in Italy in March of 2006. Teams from twenty-two countries competed; first place was won by an accountant from the Czech Republic.

The first magic square goes back much further than sudoku. Chinese literature from 2800 BC tells the story of the great flood of the Luo River. Sacrifices were made to the river god, in the hope of turning the flood. After each sacrifice, a turtle crawled out from the river. Finally, a child noticed a pattern on the turtle's back, and numbers that added up to fifteen—the number of sacrifices the river god needed to quiet the waters. Fifteen is also the number of days in each of the twenty-four cycles of the Chinese solar year.

The pattern on the turtle's shell doubles as a magic square and is also an elaborate quincunx, the pattern we read about in the chapter about five, here with three marks at the top and bottom and to either side of the important middle—a number five if the square is to add up to fifteen.

A magic square dating to the first century AD was found in the ruins of Pompeii. There's one in an inscription in Khajuraho, India, that dates from the eleventh or twelfth century. That one is a bit more complex: in addition to the rows, columns, and diagonals, the broken diagonals all have the same sum. In ancient Egypt, magic squares were engraved on stone or metal, and sometimes were worn as charms to ensure a long and healthy life. There's a magic square in Albrecht Dürer's engraving *Melancholia I.* His square totals not 15 but 34, and it has four rows across and down; the total of 34 can be found in the rows, columns, diagonals, each of the quadrants, the four corner boxes, and several other positions as well. The two numbers in the middle of the bottom row (15 and 14) are the date of the engraving: 1514. (The other two numbers in that row are 4 in the bottom left box and 1 in the bottom right box. Total for the whole row is the required 34.)

There are many other magic squares—there's one engraved on the Passion facade of the unfinished Gaudi cathedral in Barcelona. Like the Dürer square, it has four boxes across and down; they total 33, the age of Jesus at the time of the Passion. Benjamin Franklin designed a magic square that was eight rows across and down; the numbers total 260, and it has inner variations—half of each row totals half of 260, for instance.

Someone has designed a magic square made up only of prime

numbers (17, 89, and 71 across in the top row; 113, 59, and 5 in the middle row; and 47, 29, and 101 in the bottom row; the totals are 177). There are magic squares based on multiplication rather than addition, and there are also magic cubes, magic stars, and magic hexagons.

Magic squares could be made of letters, and when they were, they spelled words in the form of a cross, intersecting in the middle square. They were once considered to be religious symbols, and had particular significance in Arab countries and India. Some Islamic magic squares were believed to contain nine letters that had been revealed to Adam. The numbers in the magic square discussed in the chapter about six equaled the numerical value of the letters in Allah's name.

When the religious meanings of magic squares began to fade, the squares became superstitions—lucky charms that were even used to forecast the future. In nineteenth-century Europe and America, they were considered to be protection against fire, sickness, and disaster. Eventually, they became just puzzles, like sudoku in their way, and finally problems in a number theory.

## ANGELS AND THE POWERS OF NINE

Angels too believe in nine. There are supposed to be nine orders (groups) of angels surrounding God, arranged in triads—three groups of three. They are seraphim, cherubim, thrones, dominions, virtues, powers, principalities, archangels, and angels. Together, they are the heavenly host, and in various ways and to varying degrees, they are part of the Jewish, Christian, and Muslim religions, and several others as well.

> A tower nine stories high is built from a small heap of earth. A journey of a thousand miles starts in front of your feet.
>
> **LAO-TZU,**
> SIXTH CENTURY BC

The word *angel* is derived from the Greek *angelos* and the Latin *angelus,* "messenger," and all angels are God's messengers. They're mentioned in both the Old and New Testaments and in the Koran, with varying descriptions. In Isaiah, the seraphim are described as having six wings: two cover their faces, two cover their feet, and the last two are used for flying. In Ezekiel, thrones are said to have four wings joined together. Angels are different colors, from white to burnished brass to crystal. Some fly through the air; the sound of their wings is "like the noise of great waters." Some are invisible, but sometimes they look like extraordinarily beautiful human beings. They are pure and as bright as heaven itself; in Job, they are encompassed by light. Their number is huge.

For the Pythagoreans—for whom numbers were the music of the universe and everything could be told in numbers—man was considered to be a full chord, made up of eight notes. One more note, for a total of nine, equaled the deity. In another sense, three represented perfect unity; twice three, perfect duality; and thrice three, perfect plurality. One was the union, but three was the first number to touch perfection, and nine is three times three. Nine is the Trinity of Trinities.

> Nine-tenths of wisdom is being wise in time.
>
> THEODORE ROOSEVELT, IN A SPEECH, JUNE 14, 1917, IN LINCOLN, NEBRASKA

The powers of nine exist even aside from the potency of three times three. There are nine openings to our bodies. In Sanskrit, *chhidra* is a synonym for nine; it means "orifices," and it refers to the human body.

There used to be nine planets, but in the summer of 2006, members of the International Astronomical Union voted to change the definition of what a planet is. Under the adopted terms, the orb that has long been known as our sun's ninth planet—Pluto—is no longer a full-fledged planet; nor is it a planetoid, a planetino, a planette, or a plutoid. Officially, Pluto is now a "dwarf planet," fitting two of the new definitions of a planet—orbiting the sun and big enough to have become spherical through the action of gravity, but not, alas, the third, that it should be big enough to have cleared its nearby area of potential rivals. (There are other outer-solar-system objects: Varuna, Quaoar, Sedna, and Eris—formerly known as Zena, or more formally 2003 UB313—plus an asteroid, Ceres. There are more objects, asteroids, and dwarf planets out there; they're on a list, but they aren't yet official.) By vote, the astronomers declared that Pluto is now the prototype for a new category of "trans-Neptunian" objects, but—sadly—they voted down the possibility of calling these objects "Plutonians." Since the vote on Pluto's status as a planet, though, there have been all sorts of rumblings in the science community. Petitions are circulating, signed by NASA scientists, astronomers at major observatories, university professors, and graduate students. (Only scientists attending the summer conference were allowed to vote on the nonplanet motion.) Another international conference, not under the auspices of the IAU, is scheduled for 2007, and Pluto is on the agenda. Keep tuned.

For the record, Pluto, discovered in 1930, is 1,600 miles in diameter (smaller than Earth's moon) and crosses the orbit of Neptune on its 248-year trip around the sun. Also for the record, Walt Disney's dog was named after the planet, not the other way around. An eleven-year-old girl in Britain, Venetia Burney (now Phair) suggested naming the planet after the Roman god of the underworld, whose Greek counterpart is Hades. Her grandfather forwarded her suggestion to an astronomer who in turn sent it to the ex-planet's discoverer, Clyde Tombaugh, at the Lowell Observatory in Arizona. When the name

> At my age, I've been largely indifferent to [the] debate; though I suppose I would prefer it to remain a planet.
>
> VENETIA BURNEY PHAIR, WHO GAVE PLUTO ITS NAME WHEN SHE WAS ELEVEN YEARS OLD, AS QUOTED BY BBC NEWS, ON JANUARY 13, 2006, ABOUT THE POSSIBILITY OF PLUTO NO LONGER BEING CLASSIFIED AS A PLANET.

was officially adopted, on May 1, 1930, Venetia's grandfather gave her a five-pound note as a reward.

If you're old enough to read this book, you're old enough to think of Pluto as not only the Walt Disney planet, but also the one farthest from the sun. You may have learned a mnemonic to remember the planets in order of their distance from the sun: "My Very Educated Mother Just Showed Us Nine Planets," or "Mary's Violet Eyes Made John Stay Up Nights Proposing." Thus, the planets we grew up with are Mercury, Venus, Earth, Mars, Jupiter, Saturn, Uranus, Neptune, and Pluto. My Very Excellent Mother Just Sent Us Nine Pizzas.

Early astronomers, who couldn't see all nine planets, still believed in nine spheres: those of the moon, the sun, Mercury, Venus, Mars, Jupiter, Saturn, the firmament (the fixed, visible stars), and the crystalline sphere, which they believed revolved, carrying the stars, sun, planets, and moon with it. They didn't believe the earth could move—how could it possibly move? Birds move away from the ground, and they don't leave the earth behind them and it doesn't go ahead of them either. Birds fly away and then they return to where they started. And if the earth moved, how would we stay in place on it? The earth couldn't be a planet like the ones they could see; rather, it was the unmoving center of everything, the sun and moon, the planets and the stars—they all rotated around the earth. (I think of sitting in a planetarium, one chair amid a vast dark space, and watching the stars move across the roof of the sky. The vision of the sky moves, but the chair stays still.)

According to Greek mythology, Zeus, the king of the gods, and Mnemosyne, the goddess of memory, had nine daughters—the Muses, nine goddesses who inspire the arts and sciences. Calliope is the Muse of epic poetry; Clio, of history; Erato, of lyrics and love poetry. The Muse of tragedy is Melpomene; Polyhymnia is the Muse of sacred poetry; and Thalia is the Muse of comedy.

Urania is the Muse of astronomy; Euterpe of music; and Terpsichore of dance.

Mythology also has it that the Styx, hell's river of hate, flowed nine times around the infernal regions. In *Paradise Lost,* Milton wrote of the gates of hell as being "thrice three-fold; three folds are brass, three are iron, three of adamantine rock."

In Greek, nine is *ennea.* An ennead means nine, and in Egyptian mythology there were many important enneads, but the nine great Osirian gods (Atum, Shu, Tefnut, Geb, Nuit, Osiris, Isis, Set, and Nephthys) were the Great Ennead. All of the pharaohs and many of the gods were descended from Osiris, Isis, Set, and Nephthys. Egyptian mythology includes other groups of nine—the Lesser Ennead, the Dual Ennead, and even the Seven Enneads. Some pharaohs, few of whom were known for their modesty, created enneads of which they themselves were a part. Seti I, for one, worshipped an ennead with six deities and three deified forms of himself.

## TO THE NINES

There remain enneads of enneads of nines. The United States Supreme Court has nine justices. John Roberts is the chief justice as I write, and the other members are, alphabetically, Samuel Alito, Stephen Breyer, Ruth Bader Ginsburg, Anthony Kennedy, Antonin Scalia, David Souter, John Paul Stevens, and Clarence Thomas. As the highest judicial body in the United States, the Supreme Court leads the judicial branch of the federal government. Its members are appointed for life by the president with the

> We're all eccentrics. We're nine prima donnas.
>
> HARRY A. BLACKMUN, SUPREME COURT JUSTICE, QUOTED IN *TIME* MAGAZINE, FEBRUARY 6, 1984

"advice and consent" of the Senate. The Supreme Court is the only court mentioned in the Constitution; the other federal courts were established by Congress.

In China, nine is considered a lucky number because it has the same sound as a word that means everlasting. But in Japan nine, like the English thirteen, is bad luck. Like four, it is close to a sound that doesn't bode well—the word for nine in the Sino-Japanese system is *ku,* the same sound as the word for pain. In many Asian countries, ill health can be attributed to spirits of evil; they have poisoned breath, and are best avoided. In order to keep the spirits at bay, the Japanese avoid the use of *ku* for nine; instead they say *kokono*. But nine remains an unlucky number.

In a way, nine is also unlucky in music, because so many composers either don't finish their ninth symphonies, or never go on to write a tenth. Gustav Mahler tried to avoid the curse of the ninth by calling his ninth symphony a song cycle—*Das Lied von der Erde.* He called his next symphony the Tenth, but as luck would have it, he died before he could complete what would *really* have been the tenth, even though he called it the Eleventh—you can't fool the number god. Antonín Dvořák played with numbers too; he thought of his *New World Symphony* (he was living in Manhattan when he wrote it) as his eighth, because the score of an early C-minor symphony had been lost. The *New World* is now considered his ninth, and it is his last. Alexander Glazunov wrote only one movement of what would have been his ninth—he lived for another twenty-six years, but he never finished his Ninth Symphony. Anton Bruckner didn't finish his Ninth Symphony either, though he left sketches for the final movement. Schubert's Ninth was also his last, though he too left sketches, piano drafts, written just before his death. Beethoven's last complete symphony is—yes—the towering Ninth. (The Beethoven, Bruckner, and Glazunov Ninth Symphonies are all in the key of D minor. D for deus? dominus? death? Make of it what you will.) For the most part, the curse of the ninth is a nineteenth-century phenomenon; several twentieth-century composers—Dmitri Shostakovich among them—wrote a Ninth and survived to write more.

Unlike symphonies, cats have nine lives, testifying to the magic of cats as well as of nine. For those who don't like cats, there's the dread cat-o'-nine-tails, a whip made of nine knotted cords, believed to be a trinity of trinities in the punishment sphere, and therefore both more sacred—for those who need to justify using them—and more efficacious. Actually, the whip might have been called just a cat (no nine, no tails) because the ropes on a ship were sometimes called cats, and the whip could have been made from them. Another theory is that the whip leaves scratches like those made by a cat. A tiger or a leopard, maybe. Not your average domestic shorthair.

> **It has been the providence of Nature to give this creature nine lives instead of one.**
>
> BIDPAI, "THE GREEDY AND AMBITIOUS CAT," FABLE 3, *THE FABLES OF PILPAY,* C. 326 BC

Cats and dogs take about nine weeks to hatch their kittens and puppies. In general, the larger the animal, the longer the gestation period—and, for that matter, the longer the life. Hamsters and mice are pregnant for just over two weeks, which is partly why there are so many of them; rabbits, as Fibonacci noted, gestate for thirty-one days. People come back to nine, but months, of course, rather than weeks. And then there are whales—pregnant for an average of one whole year, and elephants, who take even longer: about twenty-two months. That's a long time to make a baby.

And then there are games. Golf has eighteen holes—two times nine. Baseball has nine innings, and nine on a team. But baseball is numerical madness all around: three strikes, four balls, three outs, double and triple plays, four bases, seventh-inning stretch, twenty-seven outs for a perfect game. In one of his old telephone routines, Bob Newhart listened on the telephone as someone described an unknown game (baseball) to him for the first time. "This is a joke," he said, "right? It's the guys in the office, right?"

Cloud nine is a state of bliss—something you might feel in seventh heaven. The usual explanation for the origin of the phrase is that it comes from the United States Weather Bureau's numbered listing of clouds. Level nine was the high cumulonimbus clouds—the big, white, fluffy clouds that look like glorious mountains in the sky—to be on cloud nine would be the height of existence. In *World Wide Words,* his online newsletter about the origins of words and phrases, Michael Quinion adds that the term also has chemical associations (a cloudy kind of euphoria), but that he suspects it evolved simply because nine is traditionally a lucky number.

> **I've been riding on Cloud Nine since the election, and I don't think I'll ever come down. Today, everything is complete.**
>
> JACKIE ROBINSON, ON BECOMING THE FIRST AFRICAN AMERICAN PLAYER ELECTED TO THE BASEBALL HALL OF FAME, IN 1962

A nine days' wonder is a brief sensation. An old proverb has it that "a wonder lasts nine days, and then the puppy's eyes are open," reminding us—should we be in a state of wonder—both that puppies are born blind and that eventually we'll see more clearly. The math of a nine days' wonder goes like this: the first three days are for amazement; the second three are to discuss the details; and the final three are for the wonder to subside and disappear.

Possession is nine points of the law. And success in a lawsuit demands nine elements, defining the nine points somewhat differently. They are: (1) a good deal of money; (2) a good deal of patience; (3) a good cause; (4) a good lawyer; (5) a good counsel; (6) good witnesses; (7) a good jury; (8) a good judge; and (9) good luck.

Someone who is dressed to the nines is as dressed up as one

can be. A tailor—not the dress-making kind—or, sometimes, a teller (who tells the news), is the stroke of a bell, the toll rung to signify a death: three for a child, six for a woman, and nine for a man. Dorothy Sayers wrote about church bells in her Lord Peter Wimsey mystery *The Nine Tailors.*

## FEMININE NINE

A novena is a recitation of prayers and devotions lasting nine consecutive days. The word is the feminine form of the medieval Latin word *novnus,* "nine each," from *novem,* "nine." Another female nine word is *novercal,* which means of or relating to a stepmother, or stepmotherly. It's not a word you'd use too often, but it's rather nice—pronounced NO-vur-kuhl. It's from the Latin *novercalis,* which itself is from *noverca,* "stepmother"—and ultimately from the Indo-European root *newo,* or "new" (the new mother), which brings us back to nine, the new number. I like thinking of nine—that masculine number, with its ball atop its straight line, a perfect outline, or a perfect sperm—as a word buried in fem*inine.*

> **There are nine hundred and ninety-nine patrons of virtue to one virtuous man.**
>
> HENRY DAVID THOREAU, "ON THE DUTY OF CIVIL DISOBEDIENCE"

Nine has always looked as it does now, with only a few changes along the way. The Indians wrote it so that it looked a bit like a modern question mark, but without the dot at the bottom. The top loop was open. It morphed into a sort of three look-alike, with its tail wrapped around itself, like the a inside of @. Eventually, the circular line became larger, and the line continued on its path downward, and the three part became smaller, to disappear finally into a tiny squiggle and then to vanish altogether. The

Arabs connected the vanished squiggle line—the top loop—to the middle of the tail, and 9 was a finished number.

> **Dreams are the wanderings of the spirit though all nine heavens and all nine earths.**
>
> **LU YEN,**
> C. AD 800

When all is said and done and finished about nine—and there is much still left untold here—perhaps the most important thing to be known about the trinity of trinities is simply this: it stands solidly as the last single-digit number. It is the gateway to ten, and all the wonders that lie beyond.

# 0

## ZERO IS NOT NOTHING

There was once a world without zero. Not a world without nothing—which would be the world before it was—but a world in which the concept of zero as a number didn't exist. The Greeks—even Pythagoras—had no zero; just their alphabet and a few borrowed letters, all of which doubled for numbers, but none of which stood for the enigmatic zero, the big 0. Neither is there a Roman numeral for zero. The idea simply hadn't happened yet. Zero had yet to be discovered. It was there all along, but it needed to be found.

> **If you look at zero you see nothing; but look through it and you will see the world.**
>
> **ROBERT KAPLAN,** *THE NOTHING THAT IS*

Most early civilizations had what are called nonpositional number systems. They used symbols in a row to show numbers, and large numbers used different symbols. The Romans could write 2007 with M for thousand and VII for seven. They didn't bother showing that MMVII had no hundreds and no tens. There was no C in that number, no X. And there certainly was no 0.

Zero came late. And slowly. Those few cultures that did have a zero in their list of numbers didn't manage to persuade anybody else, if they tried. For the rest of the world—*all* of the rest of the world—zero just wasn't there for a very long time. Zero was one of the last numbers to be thought of—and it made all the numbers that came after it, after it appeared in 10, especially the really big ones, possible. It certainly came before our contemporary googol, which is almost an exercise in zero—about which more in its turn. Before googol, which is mind-bogglingly large, came zero, which is not even small. It simply *isn't.*

Zero is much ado about nothing. It is strange contradictions and enigmas, a perfectly poetic number, everything and nothing. Most obviously, it *is* and it *isn't.* It tells you what isn't there. The question seems unnecessary, but how can nothing be a number? It isn't a quantity; it's the absence of quantity, so how can it have a place as a number?

The first counting, as we've seen, was finger counting. There is no finger that means zero, unless you hide your hands behind your back. No notch on a stick, no pebble in a pile, can represent zero—and none needs to. If you don't have *any,* you don't need to count it. If you don't have any, it isn't there. But zero *is* a number. It is an invaluable number. It is probably the single most important number—unless it ties with one for that distinction. But alone, zero does what no other number can do, not even one.

Every number—including zero—can answer the question *how many?* Look around you. Carefully. How many lions are there in the room with you? Three? Twenty-four? Oh, none, you say. Do you mean zero? You don't need zero to count lions if there are twenty-four standing next to you, assuming you're alive to count them. But if there are none, not a single one, you need a way to say that. Zero answers the question: *how many?*

> One suffices to derive all out of nothing.
>
> LEIBNIZ

The wonderful and unique thing that zero does is different from answering that basic question. What it does is clarify all the other numbers. Zero gives them a position, a home, a place to hang their hats. And at the same time it shows, when necessary, that no one is home—nothing is there. Not even a hat.

alone. In other places there are other synonyms: null, naught, or aught. Same thing. Or, if you prefer, same nothing.

We're equally uncomfortable, in a way, about infinity. We ourselves are finite, and that's hard enough to imagine. Do we have to think about endless numbers of things? We won't be here to see them or count them, and there are so many of whatever they are—drops of rain, leaves of grass, stars in the sky, empty miles of space—that in the end, numbers like that are not only unthinkable, but also uncomfortable. Maybe zero is like that, in an opposite sort of way.

> I have tried to know absolutely nothing about a great many things, and I have succeeded fairly well.
>
> ROBERT BENCHLEY

Perhaps it's the contradiction, the zero that means nothing yet makes everything possible. Perhaps it's the denial: no, there's nothing here, so don't bother thinking about it. That makes it like the elephant in the room—if we pretend that nothing's here, maybe it won't *be* here. It stays in the room—after all, it's huge—but we can pretend it isn't here because it's invisible. It's nothing.

Is that it, the concept of nothing, that makes us uneasy? Do we confuse nothing with nonexistence? Zero is the absence of anything, but whatever it is isn't gone forever; it's just not here right now. It could be right back. That's different from not existing at all, but maybe we confuse the two concepts and we'd rather stay altogether away from the idea.

Maybe it's just that zero confuses us. Early computers put a slash through zero so it would look different from the letter O. Some computers still use that slash, but it sets up its own confusion because the Scandinavian vowel is also an O with a slash through it. The zero-slash doesn't go beyond the walls of the zero, which distinguishes it from the Scandinavian—there the slash thinks outside the box—or if you prefer, the circle. But the difference is slight, and the two symbols are still confusing. The line across the middle of 7 is better at eliminating confusion, this time with 1, without trespassing on a letter in somebody else's alphabet. Even beyond the Scandinavian letter, Germans use a slashed letter O to mean average, or average value, *Durchschnitt*, literally "cut-through." The letter O is rounder and fatter than zero, a circle as opposed to an oval—O against 0—but when they're not together, they can be hard to distinguish. Still, when we're speaking and not writing, there is no confusion in saying "zero" instead of "oh," and we still tend to avoid it.

We avoid using zeros in buildings as well as in speech. There is no zero floor. Elevator buttons never say "0," only "Lobby." In the United States, there is only the ground floor—one flight up is the second floor. In many other countries, one flight up is the first floor. Does a zero floor not exist anywhere? How clarifying it would be to have one! Hossein Arsham, at the University of Baltimore, notes that the Department of Mathematics building at the University of Zagreb has numbered its floors in a perfectly logical way. The basement is −1, minus one. The ground-level floor is zero. Up one flight is the first floor. Some buildings in Spain and in Latin America do the same thing—ground level is called zero. One is one flight up. Arsham suggests that one way of explaining the Spanish willingness to use zero as a floor name is that "Islamic culture had more influence in Spain than in any other European country." Islamic countries had zero way before European countries did. After first appearing in India in the second or third century BC, it had traveled to a caliph's court in Baghdad by the eighth century AD. It appeared in print for the first time in 820, in an arithmetic textbook by al-Khwarizmi, from whose name we have the word *algorithm*. Muslims brought zero to Spain early in the twelfth century, and Fibonacci's *Liber abaci* was published in 1228. That's a long time to wait for zero—a millennium and a half.

Zero does have negative connotations. If a team doesn't score any goals in a hockey game, their score is zero—a goose egg. (In tennis, zero is called love, which sounds quite nice until you find out that this *love* is from the French *l'œuf*—"the egg." In British cricket, *duck's egg* is used in the same way Americans use *goose egg*. A duck's egg—or just a duck—means a batsman is out for a score of zero—one way not to score—and it's from the way zero looks in a score book. Like a duck's egg.) If you don't answer any questions correctly on a test—not one—zero again, usually in red ink. You failed! If you have no money in the bank, your balance is zero, and if you write a check anyway, it'll bounce, and you'll be in negative numbers. There are many other unpleasant or difficult zeros, and we'll come to some of them, but the idea is clear: zero makes us uncomfortable. No other number can say that.

> **For nothing, you get nothing.**
>
> **MY MOTHER**

Is zero the end? Or is it the beginning? Do we count 1-2-3-4-5-6-7-8-9-0? or 0-1-2-3-4-5-6-7-8-9? The computer keypad, and the typewriter before it, put zero at the end of the numbers on the top row. The telephone puts it after nine, too, on the bottom—but in a place of honor, since it, not O, as in MNO, stands for operator as well as for zero.

You do need a zero before you can write a ten; we go from 9 to 10, and we need the zero in order to do that. Is zero odd, then, or even? Arsham tells the story of a 1977 smog alarm in Paris, as reported on German television. Only cars with license plate numbers ending in odd numbers were to be allowed to drive in the city, in order to cut down on pollution. Cars with license plates ending in even numbers were temporarily illegal. But what would happen to cars with license plates ending in zero? As it turned out, no fines were issued to those cars, because the police didn't know whether to call them odd or even.

There are those, like the Paris police, who say that zero has neither property; it is neither odd nor even; it simply *is*. Others say it's even because it comes before one and after nine, and odd and even always alternate. It definitely fails the test of dividing evenly in half. Zero is incomplete unto itself; you can't divide zero. No point in trying.

Zero, in its way, is a gateway number. It divides positive from negative numbers; on the one side are the pluses; on the other, the minuses. Zero made it possible for mathematicians to think about negative numbers. By itself, zero is neither positive nor negative. It just is. Or isn't. It's like "Go"—it has to be passed. But you can't buy it, and you don't have to pay rent when you're on it. Zero is always just not there.

> When one's expectations are reduced to zero, one really appreciates everything one does have.
>
> STEPHEN HAWKING

On both the Fahrenheit and Celsius thermometers, zero divides possibilities; it separates. Zero on a Celsius thermometer is the point at which water freezes; below zero, there's ice. On a Fahrenheit thermometer water freezes at thirty-two degrees; zero on that scale is a sort of minus number—thirty-two degrees below the freezing point.

Zero is also a co-dependent number. It can stand alone, but it probably gets a headache after a while. It is barely acknowledged when it passes by—it's Charlie Brown with Lucy always pulling the football away. If you put zero before a number, nothing much happens: 1 and 01 (without a decimal point) mean more or less the same thing. But add zero to the end of a number, and you really have something. Change 1 to 10, and you've multiplied by ten, turning one into something considerably larger. Zero did that! To the right of a decimal point, zero is a big shot in a different way. Point-oh-one is considerably smaller than point-one (.01 versus .1)—ten times less, just as, in the other direction, 10 is ten times more than 1. The power of zero lies in being in the right place at the right time, and zero does that extremely well.

## THE BIRTH OF ZERO

Fingers, piles of pebbles, notched sticks, differently shaped symbols, letters that in some places doubled as numbers—these were all logical steps, using simple and available materials and answering the basic mathematical questions. Sand or dust tables used the same principles of ease and availability but were just a bit more complicated.

At first, a dust table was a drift of sand on the ground on which figures could be scratched with a finger or a stick; eventually, the sand was on a table. Sand tables had inherent in their structure an idea of place value. They made clearly visible a progression of numbers, in columns—the single digits, the tens, the hundreds, the thousands. . . . The thing is that nobody before the Hindus in India—or for a long time after—noticed that; or they noticed but they just didn't make the leap from sand table to basic idea, to *writing*.

By itself, the sand table was an advance; it made more rapid calculations possible. But, like so many other things before, it was temporary. A smudge could be a disaster. Taking a sand table with you was out of the question—you had to find a new space, a new handful of dust. The abacus was a more functional translation of the sand table, portable and relatively permanent. It could be large or small, public or private. Beads are strung along its wires to represent columns of numbers—the single numbers one through nine, the sets of ten, the hundreds. When there is no number to put in a column—all the beads have been moved, and no new ones need to be slid into place—there's an empty space. The Sanskrit word for that empty space was

> Even death is unreliable. Instead of zero it may be some ghastly hallucination, such as the square root of minus one.
>
> SAMUEL BECKETT

*shunya,* meaning "empty," "void," or "blank." When it was written as part of a permanent record of counting-board work, *shunya* was exactly what it said—a circle around an empty space: 0.

In India, for the first time, a series of ideas had now been combined that made possible the number system we know today. There had been a series of steps, each simultaneously small and huge. The Hindus used abstract numbers that weren't written as a series of lines to be counted, but instead were nine separate and distinct symbols that didn't double as anything else. They didn't invent new symbols for higher numbers; only those nine numbers were used to write every number they wished to write. They used a placement system, so that the value of a number could be determined by its place in a row of numbers. They learned from the sand table and the abacus, and then went further to *write* numbers in the same column order they saw on the wires. They understood that they could use an empty space to indicate a place not filled by a numeral, and finally, they replaced the empty space with a new symbol, a sign that could be written, a new number: zero.

## SEEING THE VOID

It's hard for us to imagine counting without that *shunya,* that nothing, that zero. Without it, how could you write a number higher than nine? You could have a word that meant ten, or a symbol—as the Romans did, X—but in order to quickly, cleanly, and clearly write and read numbers above nine (or above ninety-nine, or above nine hundred ninety-nine), you'd need zero. Small as it is, that vacant circle is mighty.

Zero is more than vacant, though. It's a void, it's emptiness. And the Hindus and Buddhists had complicated concepts of the void before the rest of the world did. It's possible to list twenty-five different kinds of *shunya*—an exercise something like the listing of words that define subtle differences in snow. For *shunya,* Georges Ifrah (who listed the ideas symbolized by infinity's serpent figure) lists "the void of non-existence, of non-being, of the unformed, of the unborn, of the non-product, of the uncreated or the non-present; the void of the non-substance, of the unthought, of immateriality or in-substantiality; the void of non-value, of the absent, of the insignificant, of little value, of no value," among others. Hindu mathematicians used as synonyms for *shunya* words that meant sky or atmosphere, space, the firmament—the stars and heaven. *Shunya* is the Serpent of Eternity, the infinite, the void, and it is also nothing.

> **The less anything is, the less we know it: how invisible, how unintelligible a thing, then, is this Nothing!**
>
> **JOHN DONNE**

Infinity's sleeping eight, with its two loops, could very well be two zeros, touching, and zero is a kind of infinity—it's an endless nothing. There are still, in many languages, a host of synonyms for *zero,* just as there are for *many,* and *many* is another, more manageable, way of thinking of infinity. Among zero's names are nothing, nil, nada, aught, cipher, goose egg, naught, zilch, zip, null, zero, and plain old O, as in oh.

In Sanskrit, there's also another term for zero: *bindu. Bindu* means "dot," and it was used because sometimes the circle of zero was condensed into a dot. *Bindu* is a specific space, unlike *shunya,* which is an empty space. A dot is the center—the center of the center, the core of the circle, the inside of the eye. For Hindus, *bindu* was related to a representation of the universe before creation—a concentration of energy so intense that it was capable of creating everything. That dot, that uncreated universe—a reverse black hole, not an annihilation of everything, but instead the possibility of all—was at the same time a link to the void, and both are circles. A circle was a logical symbol for an empty space which makes so many things possible. The circle in this sense is not unlike the color black, which holds all colors. Making the leap from zero to infinity was another concept which came logically to the Hindus, who believed in everything and nothing.

## WESTWARD O!

*Shunya*—and the accompanying numbers from one to nine—traveled slowly to the west, from India to the Arab countries, where mathematicians translated *shunya* to the Arabic word *sifr,* still meaning empty space.

Fibonacci, in his book *Liber abaci,* called the Arab *sifr* by a new name: *zephirum,* and that name was then used in Italy. The word was a natural transition. In Latin, *zephyrus* is the west wind, a light breeze, what we would call a zephyr, a light breath of wind—almost nothing.

Not long after, *sifr* arrived in Germany, where it was translated into a sound-alike, *cifra.* Even with all this translation, *sifr, zephirum, cifra,* very few people understood yet about zero and what it could do. As a result, *cifra* gradually came to have the meaning of a secret sign. From *cifra* comes *cipher*—code. And *decipher*—to figure out.

*Cifra* eventually also became the basis of cipher in a different sense. Because most people recognized the new Arabic numbers by their newest and most important feature—zero—they called *all* the new numerals by that single number's name, *cifra, ziffer,* or *chiffre,* depending on where they lived. And because most people were using *cifra,* or its transliterations, to mean all the numerals and how they were used (in English, *cipher* and *ciphering*), mathematicians and other scholars were left needing a word to mean specifically the number zero, rather than the numerals as a whole. They adopted the Italian *zephirum,* and we ended up with *zero.*

In some languages (German and Old English), the word for

zero is *null. Null* is a direct descendant of the Latin phrase *nulla figura*—figure of nothing, no number, no *thing* at all. Null and void. Emptiness. A circle around an empty space. Nothing there. Void. Zero.

## THE BATTLE AS IT WAS FOUGHT

Counting boards existed in many places—they traveled easily. Greeks, Romans, Chinese, Indians—all, and others, used counting boards to do their arithmetic. The etymology of the word *abacus* shows a bit of its history: it's a Latin word, descended from the Greek *abax,* "table," which itself is probably *avaq,* from either the Hebrew word for dust, or *abak,* the Phoenician word for sand, all related to the sand table. There are still many places today where fingers fly over an abacus, beads clicking as they're moved from place to place. It's really been only a relatively short time since the abacus was replaced by pencil and paper (and eraser)—and the mere flicker of an eyelid since the adding machine and then the electronic calculator took over the hard work of arithmetic. *Calculator* is from the Latin *calculus*—"pebble"—as in both counting by pebbles and the beads strung on the wires of the abacus. The Latin verb *calculare* means literally "to pebble" and figuratively "to do arithmetic"—thus *calculus, calculate,* and *calculator.*

Long before the calculator, the numerals we know now were emerging from India and, as word of their arithmetic wonders spread from place to place, traveling west together. In a perfect world, we'd call our numbers Indian numerals. But we call them Arabic because we learned about them from Arab scholars, who had learned in turn from Indians.

It seems possible that Fibonacci used the word *abacus* in the title of his book, *Liber abaci,* in order to avoid antagonizing the accountants and professional mathematicians who were accustomed to using the abacus and who had little or no knowledge of the new numerals, less interest in learning about them, and zero willingness for anybody else to use them.

Their numerical knowledge put them in a separate category from most people. Their work was an arcane art; it seemed magical to those who didn't know how to manipulate numbers, and it gave them power. Many were clerics and monks under the protection of the church. They were certainly inclined to keep what they knew to themselves, and they saw the new numbers as a threat—the numerals were so easy to work with; anybody could learn what to do with them. What would happen to their power then?

The church had its own interest in maintaining the same kind of mystery. Science and philosophy were matters for the church to control, and they were supposed to support, not differ from, religious dogma. (It was as late as the seventeenth century [in 1633] that Galileo was put under house arrest by the church and publication of his works was banned because he advocated the Copernican theory that the earth traveled around the sun, making the sun, rather than the earth, the center of the known universe.)

Church control meant the narrowing of learning rather than its expansion. According to Georges Ifrah, "Some ecclesiastical authorities thus put it about that arithmetic in the Arabic manner, precisely because it was so easy and ingenious, reeked of magic and of the diabolical: it must have come from Satan himself!"

No wonder Fibonacci was careful! The churchly mathematicians became known as the Abacists because they continued to use the abacus. They did their calculations on the abacus and they wrote their answers in Roman numerals—holding the abacus with one hand and writing MCCCXVII with the other. (In medieval Italy, *abaco* was the normal term for written arithmetic.)

The battle between the Abacists and the Algorists (those who gave up the stateliness of Roman numerals for the ease of the Arab numbers) lasted for centuries. Along the way, there were places where the new numbers were against the law. Even aside from the church's position, it was thought that they tempted forgers because they were more easily falsified than Roman numerals could be. Arabic numerals were banned in banks and were not allowed to be used on official documents. Algorists were sometimes even burned at the stake, as if they were witches or heretics.

> **What is man in nature? Nothing in relation to the infinite, all in relation to nothing, a mean between nothing and everything.**
>
> BLAISE PASCAL, *PENSEES*

But the result of this prolonged struggle was inevitable. The numerals became a kind of secret code (yes, a cipher), used by merchants and businesspeople who were willing to evade the laws and the secret arts—after all, the numbers were there, and they were fast and easy to use. Finally, by about the beginning of the sixteenth century, they were here to stay, though there were still those who double-checked their computations on an abacus just to be sure. (There are still many places where the abacus is preferred to the computer or a calculator because the work done on either of those isn't visible, while the computations worked out on an abacus can be seen by anyone who cares to watch.)

In the end the numerals were irresistible. The Algorists had won. By the eighteenth century, the abacus had virtually disappeared from Western Europe, and 1, 2, 3, 4, 5, 6, 7, 8, 9, and 0 looked very much as they do today.

Those numbers are elegant in their simplicity and versatility. There are only ten of them, but those ten can make billions.

## PERFECT ZEROS

There are famous zeros. If you're old enough, you may remember the Japanese warplanes of World War II. There's ground

> At zero minus one minute, all observers at Base Camp, about 150 of the "Who's Who" in science and the armed forces, lay down prone on the ground in their preassigned trenches, face and eyes directed toward the ground and with the head away from Zero.
>
> **WILLIAM L. LAURENCE,** IN THE *NEW YORK TIMES,* SEPTEMBER 26, 1945

zero, a pinpointed target, and the site of the destruction of the Twin Towers on 9/11. Ground zero was originally the area above or below the detonation of a nuclear weapon, but it has come to mean the exact spot on the ground where any explosion occurs, or the center of an earthquake or hurricane or any other disaster. The farther away you are from ground zero, the less destruction there is.

There's absolute zero, not a bottle of vodka, but the temperature at which a thermodynamic system is at its lowest energy point, the point at which the motion of particles which constitute heat is minimal. It's way chilly minus several hundred degrees. Molecules are supposed to stop their eternal vibrating at this point (it's their hot jiggling that causes water to boil). Molecular motion doesn't exactly stop at that temperature; the molecules may be doing their best to wiggle even at absolute zero, but we're told that there is no energy available to be transferred from that cold place they're in to any other system. Result: energy is minimal. Theoretically, absolute zero is the lowest possible temperature. Modern physics, though, also tells us that subatomic particles never really stop wiggling altogether, so as luck would have it, absolute zero isn't really absolute. (Nothing ever is.) It's still zero, though—in fact, way below zero.

> Nothing will come of nothing.
>
> **WILLIAM SHAKESPEARE,** *KING LEAR*

Zero hour is when the sun is at its meridian—high noon. It's also the moment when something is due to happen. To zero in is to come closer; zero gravity means there is none; zero population rate growth means there is no growth going on—two parents have two babies, and in the end, to which we all must come, there is no body lost, no body gained. And zero is when liftoff happens, not at one in the countdown, but at zero: ten, nine, eight, seven, six, five, four, three, two, one, zero, ignition, liftoff. Zero is when it happens.

Zero is famous for what it demarcates: the precise place, the exact hour, the point of liftoff. In that sense, zero is the fleeting moment of now.

At birth, we are zero. Between then and our first birthday, our age is measured in months: that adorable baby is six months old, or three months old, or eight months old. A year after birth, we light one candle on the cake—the baby is a year old! Zero is exactly what was missing when time went from 1 BC to AD 1. There was no zero year to mark the change. During what we call AD 1, we were only in our first year—at the end of that year, we reached our first birthday; like the baby whose birthday is marked by a single candle, the century was one year old, and the following year should have been called 1. It's like a ruler: the one-inch mark shows the difference between zero—the end of the ruler—and one. All of this makes for controversy and conflict every thousand years or so. Purists told us that this millennium didn't actually begin until January 1, 2001, and realists said, "No way, it got here when the numbers changed from 1999 to 2000." The purists were right; that's what makes them purists. But the realists had usage on their side, and usage often counts more than being right.

## ZERO IS SOMETHING

Both as a number and as an idea, zero does strange things. The very thought was troubling even to mathematicians at first. Like the church, they thought the idea of zero was nothing more than the devil's handiwork. How could zero be a number—be *anything,*

> It can be a simple sentence that makes one single point, and you build for that. You zero in on one moment that gets that character, you go for it, that's it, man, and if you fail the whole thing is down the drain, but if you make it you hit the moon.
>
> JACK LEMMON

in fact—if what it represented was nothing? If adding zero to the end of a number multiplies it by ten, how can zero be nothing? When you add or subtract, zero tells you that it isn't there. How do you prove the existence of something that isn't there? When you multiply by zero—oh! when you multiply!—zero annihilates; it's all-consuming: five times zero is nothing; it's zero. The five is gone. Only zero—nothing—exists. When adding or subtracting, then, zero is ignored; when multiplying it conquers; when dividing, it is an impossibility at worst and an enigma at best. Zero doesn't count numbers of things; instead, it indicates the things that aren't there. That's like what Gertrude Stein said about Oakland: there's no there there.

Zero is a magic and a mystery all of its own making. It isn't a Freudian enigma, though. If the numeral one is masculine, with its straight phallic line, zero is clearly feminine—round, curved, a place that's empty and waiting to be filled. Zero—0—is an egg, an ovum, a seed, a womb, the beginning of all things. Zero is that out of which one is born. To say it differently, zero is the potential and one is the activating force. If you forget Freud: One is the unity, one is God, and God is one. Zero is the void. But zero makes everything else happen.

Thinking about zero can be dizzying, something like traveling around the circle that it is and realizing that it encloses an empty space. It's everything, and nothing. It either is or isn't—it can't be somewhat zero, any more than something can be very unique. Zero *is* unique. It's absolute. It's an absence and a presence, the symbol of emptiness and at the same time the number which stands for the absence of everything, which means that it's something. It's an empty place, but it's waiting to be filled. It isn't the opposite of one, though it makes possible the negation of one. Is it the ancestor of one? The amplification of one? Zero makes for mind games—but when zero was a new idea, a new number, these were questions that needed to be answered.

> **It's not true that I had nothing on. I had the radio on.**
>
> MARILYN MONROE, ON POSING NUDE FOR A CALENDAR

Clay is molded to make a vessel, said the Chinese philosopher Lao Tzu more than two thousand years ago. "The utility of the vessel lies in the space where there is nothing. Thus, taking advantage of what is, we recognize the utility of what is not." In the spirit of the contradictory nature of nothing, zero is the beginning and the end. It is the nothing that makes everything possible.

# 10

## TEN MEANS TWO HANDS

Reaching ten has an almost finished sort of feel—something like the click a combination lock makes when the last number fits into place. Ten: a single syllable, simple to say, easy to spell, the last of our fingers. Ten feels right.

> This number was of old held high in honor for such is the number of the fingers by which we count.
>
> OVID

For the ancient Greeks, ten not only felt right, but was a number of such importance that it was considered holy. Ten, the Pythagoreans believed, was the single number that included *all* numbers and therefore it represented the cosmos, which was made up of numbers.

Because ten is the sum of one, two, three, and four ($1+2+3+4=10$), they also believed it included the properties of those numbers. One meant unity, the single indivisible thing from which all the numbers arose. Two, with its opposites, was polarity; three, harmony. And four meant space and matter—and the four elements (earth, air, fire, and water). Thus, because unity, polarity, harmony, space, and matter were combined to form ten, it was the number of everything, everything that was, everything that is. Philolaus, one of the Pythagoreans, said, in around 400 BC, that ten was "sublime, potent and all-creating, the beginning and the guide of the divine concerning life on earth."

For the Pythagoreans ten also represented the geometric forms:

one, the point; two, the line; three, the plane; and four, the solid—and again, the four numbers combine to make ten. Ten again contains everything. Ten, they said, was "the first-born of the numbers, the mother of them all, the one that never wavers and gives the key to all things." They represented ten as a triangle of ten parts, made up of four rows, representing the order of the creation of the universe:

•<br>
• •<br>
• • •<br>
• • • •

Because they believed this triangle, which they called the Tetraktys, to be a holy representation of the number ten, they swore oaths by it, as Christians do today on the Bible. The Tetraktys (from the Greek *tetra,* four) was an important part of the initiation rites for new members of the brotherhood; novices took their oaths to the Tetraktys, and then began a three-year period of silence. "By that pure, holy, four-lettered name on high," the oath said about the Tetraktys, "nature's eternal fountain and supply, the parent of all souls that living be, by him, with faith find oath, I swear to thee."

The Pythagoreans supposedly based their musical system on the Tetraktys, reading the rows as the ratios of four to three, three to two, and two to one—the basic intervals of the Pythagorean scales.

> Bless us, divine number, thou who generates gods and men! O holy, holy Tetraktys, thou that containest the root and source of the eternally flowing creation! For the divine number begins with the profound, pure unity, until it comes to the holy four; then it begets the mother of all, the all-comprising, the all-bounding, the first-born, the never-swerving, the never-tiring holy ten, the key holder of all.
>
> PYTHAGOREAN PRAYER

Later, the Tetraktys influenced not only the early Kabbalists but also the reading of Tarot cards. Catholic archbishops' coats of arms include a design of two separate Tetraktyses.

## RAISING TEN

Whether or not ten comes from the first unity, ten is the first step toward a multiplicity—one hundred, and then a thousand, and then ten thousand, a vast number which the Greeks called a myriad.

Multiplying by ten—1, 10, 100, 1,000, 10,000—is what we do automatically because the base of our numbers—in the decimal system—is ten. If our base were a different number, we'd have different larger numbers (and new columns of numbers) in multiples of our base number. In the binary system, for instance, every time a new power of two is reached, there's a new column:

| BINARY NUMBER | DECIMAL NUMBER |
|---|---|
| 1 | 1 |
| 01 | 2 |
| 11 | 3 |
| 100 | 4 |
| 101 | 5 |
| 110 | 6 |
| 111 | 7 |
| 1000 | 8 |
| 1001 | 9 |
| 1010 | 10 |

The long string of numbers and the many columns are two of the reasons not many people have counted by twos. But they *have* counted by *other* numbers. We have many leftover examples of other ways of counting. Systems based on twelve are the easiest examples: think of twelve hours on the clock, twelve inches in a foot, or a dozen eggs; more of all of this when we reach the number twelve. We've also counted by sixteens—think of ounces in a pound—and by fives, twenties, and sixties.

The Sumerians and then the Babylonians, who came first to the concept of zero, counted by sixties. That seems less strange when we remember that we, too, count by sixties when we think of time, the degrees of angles and arcs, or the minutes and seconds for the earth's lines of latitude and longitude. The Greeks and Romans inherited this sixtyish point of view when they divided the day into twelve hours and each hour into sixty minutes.

No one knows for sure why the Sumerians and Babylonians worked with the base of sixty—though it has the obvious advantage of being divisible by many numbers: one, two, three, four, five, six, ten, twelve, fifteen, twenty, and thirty. Still, sixty seems a cumbersome number to work with as the base of a numerical system—it's so large; it probably means remembering a great many number words. But they divided the base of sixty into subsets of ten, and that helped enormously. After ten, compound numbers were used to reach sixty—just as we use them to reach hundred.

> All lovers swear more performance than they are able, and yet reserve an ability that they never perform; vowing more than the perfection of ten, and discharging less than the tenth part of one.
>
> WILLIAM SHAKESPEARE, *TROILUS AND CRESSIDA*

In English, everything between twelve and a hundred is made up of numbers that have gone before and now define the new number: eighteen is eight and ten; thirty is three tens. For the Babylonians, each number from one to ten had its own name. After that, numbers were combinations of one and ten. Until sixty, which had its own word, the first new name since ten. The word for sixty resembled the word for one; it also meant unity (something like the

Pythagorean view of ten)—probably because sixty represented the union of so many previous numbers. Beyond sixty, number names were compounds until six hundred was reached (which is sixty times ten); six hundred had its own name. Thirty-six hundred (sixty times sixty) had a new name, and the number names proceeded in that fashion. The written numerals (cuneiform symbols written on clay tablets) worked in the same way, with separate graphic symbols for one, ten, sixty, six hundred, three thousand six hundred, thirty-six thousand . . . all based on six and ten.

The Sumerians came to zero as a separator when they wrote large numbers. Scribes began leaving a blank space between symbols to indicate that nothing was there (as in 205: no tens), but the blank space was confusing, and eventually a small symbol, a sort of sideways wedge (like an empty martini glass on its side), was used instead: zero. They didn't put zero at the end of a number, though; it was assumed that whoever was working with the number knew how large it should be, or, if that might not be true, a phrase was added instead. On an accounting tablet noting distribution of grain, a row of numbers ends with the phrase "The grain is exhausted." In a description of subtracting twenty from twenty, the scribe wrote, simply, "You see." Zero was not yet a number.

Hundreds of years later—and on another continent—the Mayans, who counted by twenties, also had a zero. Time, for the Mayans, was cyclical; there were hundreds of celestial and earthly cycles of time, each kept as a separate calendar. Each calendar in this complicated system had a base period of twenty days, and the periods of time had spaces that needed to be filled in. When there was no unit for a space, the Mayans used a graphic symbol to stand for the empty place. It looked a bit like a clam with a few stripes on its top. This clammy zero wasn't considered a number; it was just a space marker, and it wasn't used in calculations.

The Aztecs too counted by twenties. Their first five numbers—the fingers of one hand—had their own names. The next five—the other hand—were combinations of five and the relevant numbers, so that six was five-and-one; seven was five-and-two (as our thirteen is three-and-ten, fourteen is four-and-ten), and so on until ten, which had its own name. The next five numbers were linked to the toes of one foot, and were ten plus the relevant number until fifteen, which had its own name; and the last five, toes on the other foot, were compounds of fifteen plus the relevant number until twenty, which had its own name. The Aztec day was divided into twenty hours. Their army divisions had eight thousand soldiers—eight thousand is twenty times twenty times twenty. Their word for twenty also meant one man—ten fingers plus ten toes.

The Banda, in central Africa, used words meaning "a hanged man" for their number twenty because when a man is hanging you can see all his fingers and toes. A tribe in Senegal used "a man" to mean twenty, and "a bed" to mean forty—which presumes that the bed is shared by two people, but omits the woman, if woman it was.

The French language shows residues of counting by twenties. The number eighty in French is *quatre-vingts* (four-twenties). Ninety is *quatre-vingt-dix,* four-twenty-and-ten. Until the seventeenth century, there were many other multiples of twenty in common use. Molière, in *Le Bourgeois gentilhomme,* uses *six-vingts,* six-twenties, for one hundred twenty. A hospital in Paris, originally built by Louis XI to serve three hundred blind veterans, is still called the Hôpital des Quinz-Vingts (fifteen-twenties, or three hundred). Danish, Basque, Breton, Welsh, Irish—all have traces of twenty-counting.

The Bible cites an average life span of three score and ten—if you're older than seventy, you're ahead of the game. In English, we have Abraham Lincoln's "Four score and seven years ago . . ." Much better than saying eighty-seven years ago—and the same lingering trace of using twenty as a base for counting.

For all of that, it's ten that comes naturally. Ten may not be the best number to count by—it doesn't have many divisors, just one,

> It is my ambition to say in ten sentences what every one else says in a book—what everyone else does not say in a book.
>
> FRIEDRICH NIETZSCHE, "SKIRMISHES OF AN UNTIMELY MAN," *TWILIGHT OF THE IDOLS*

two, five, and ten, and any number can be divided by one and by itself. But numbers are organic; our hands were our first calculating machines, and they remain our best. Our fingers give the ten numbers a feeling of exaltation—of existing on a higher plane, of being the essence of numbers.

Ten-counting is not universal, but it almost is, and it almost always was. Arabs, Egyptians, Chinese, Etruscans, Greeks, Hebrews, Indians, Romans, Incas—all used base ten. Among history's strangest ten-counters were the Incas, whose empire stretched to include parts of what is now Ecuador, Chile, Peru, Bolivia, and Argentina. The Incas had no written language, so they memorized messages that needed to be carried from the capital at Cuzco to the rest of the empire. They used relay runners along the Highway of the Sun, a system of roads with stations every four and a half miles. The Incas recorded numbers by tying knots on strands of string, a system called *quipu.* As many as a hundred strings could be bundled together in one group with each strand showing one number through different kinds of knots, tied at different distances. Calculations were made separately, and the results were recorded on the strings. There have been other knot-tying number systems, in parts of Africa, ancient Greece, China, and even in Hawaii, most often based on counting by tens.

Around the world today, ten is the common language of numbers, spoken everywhere from Mongolia to Samoa, from Romania to Thailand, from Finland to Turkey, and most places between. (Except perhaps at the Walt Disney studios, where a long time ago it was decided that Mickey Mouse and his pals should have only three fingers and a thumb on each hand, because that made the drawing of all those animation cels much easier than ten fingers would have been. "Leaving the [fifth] finger off," said Walt Disney, "was a great asset artistically and financially. Artistically, five digits are too many for a mouse. Financially, not having an extra finger in each of 45,000 drawings that make up a six-and-one-half-minute short has saved the studio millions." It seems likely, then, that Mickey counted by eights, if not by fours. Or perhaps he counted by millions.)

> Life is like a ten-speed bicycle. Most of us have gears we never use.
>
> **CHARLES M. SCHULZ,** CREATOR OF *PEANUTS*

## REGARDING TEN

Ten is not only the last finger to be counted, it's also the first of the double-digit numbers. If you count on your fingers and you want to go higher, reach for your toes. Ten is as far as you can go with your gloves off.

The word for ten, some scholars think, has its roots in an old word that meant two hands—or, by definition, ten fingers. Ten's clear roots are in the Indo-European prototype *dekm.* Sanskrit is *dasa;* Greek is *deka;* Latin is *decem.* In Italian we say *dieci,* in Spanish *diez,* in French *dix.* Swedish is *tio,* German *zehn,* Czech *deset,* Albanian *diete,* and in Anglo-Saxon *tyn*—a straight line to the English *ten.*

Out of all these words comes denary, or decimal, as the method of counting by tens; the decimal system (with the decimal point) is an extension of that, our way of dealing with percentages and fractions and pennies—a handy way, given our fingers, of systemizing things by tens.

December, originally the tenth month in the Roman calendar, and dime (shortened from *decime*), a tenth of a dollar, are from the Latin *decem.* A decade is ten years long. A dean, or a deacon (from the Latin decanus), was first the soldier in charge of a tent containing ten men, then the head of ten monks, and finally—or at least now—the head of a cathedral or a college. To decimate was originally to execute one person in every ten, thus drastically thinning the ranks. In Rome, one soldier killed in every ten was the punishment for cowardice or mutiny.

Library books are also organized by tens in a system devised by Melvil Dewey in the 1870s and known now as the Dewey decimal system. It works this way: Generalities are classified at the beginning, with 000; philosophy and psychology are in the 100s. Religion begins at 200; the social sciences at 300; language at 400; and the natural sciences and mathematics at 500. Technology and applied sciences are in the 600s; the arts begin with 700; literature and rhetoric with 800; and geography and history at 900. Each count is further divided by tens; each ten is divided by tens again; and then by a decimal point and its following numbers. Every book in the world fits somewhere in that system, and it is used by hundreds of thousands of libraries around the world.

A tithe is one-tenth of something. (It's from the Old English *teogotha,* "tenth.") Tithes are either voluntary contributions or involuntary taxes, paid to support a specific religion, usually one which is Jewish or Christian. Once, tithes could be paid in agricultural products—cattle or grain, for instance. Today, hard cash is preferred; please don't leave a cow at the church door.

## POWERS OF TEN

The Lost Tribes of Israel numbered ten: Asher, Dan, Ephraim, Gad, Issachar, Manasseh, Naphtali, Reuben, Simeon, and Zebulon. They formed the Kingdom of Israel, in the north of the promised land of Canaan, while two other tribes—Judah and Benjamin—established Judah in the south. The northern kingdom was captured by the Assyrians in 721 BC, and through a process first of forced dispersion and then of assimilation, the ten tribes disappeared into history. The belief persisted that they would eventually be found, and over the years there have been many who either were said to be, or claimed to be, one of the Lost Tribes: Afghans, Ethiopians, Native Americans, Mormons—even some Japanese.

The Ten Commandments are the laws given to Moses on Mount Sinai by God, and in turn by Moses to the people of Israel—part of the story of the Exodus from Egypt.

> **There have to be ten commandments because, if there were twelve, when the priest counts one, two, three, holding up his fingers, and comes to the last two, he'd have to borrow a hand from the sacristan.**
>
> UMBERTO ECO, *FOUCAULT'S PENDULUM*

Exodus goes on to say that after Moses saw that the Israelites had gone astray and that his brother Aaron had made the Golden Calf for them to worship, he broke the two tablets on which the Commandments were inscribed. But later Moses brought a second set of Commandments down from Mount Sinai; this was the set placed in the Ark of the Covenant, and this time, the Commandments were honored.

The Commandments are a moral foundation that in one form or another began in our early history. They are a summary of the basic shalls and shall nots that speak of human responsibilities to God, to society, to family, and to oneself—accepted by Jews, Christians, and Muslims as a summary of the behavior God expects from humanity. Almost every modern religion, whether or not it recognizes the Ten Commandments as such, includes comparable laws or principles.

For Jews and Christians, the Commandments are of paramount importance. Muslims believe Moses to be one of their greatest prophets, but Islam teaches that the biblical text has changed from the divine original over the centuries through carelessness at best or malice at worst. Muslims believe that the Koran includes the revelations of both the Old and the New Testaments, and there are specific verses in the Koran that compare to the Ten Commandments, beginning with "There is no other god but God."

> In vain we call old notions fudge,
> And bend our conscience to our dealing;
> The Ten Commandments will not budge,
> And stealing will continue stealing.
>
> JAMES RUSSELL LOWELL
> MOTTO OF THE AMERICAN COPYRIGHT LEAGUE WRITTEN IN 1885

Christians believe that the Commandment that bans worshipping graven images has more to do with the idea of worship than with the images as such, and that one may build "likenesses," as long as they are not worshipped. Many Christian churches feature images, paintings, statues, and, in Eastern Orthodoxy, icons. Jews, Muslims, and some Protestants read the Commandment differently, and believe that it prohibits the use of images and idols in any way. Catholics criticize the Orthodox veneration of icons; there are some Protestant groups that criticize the way stained-glass windows are used by other denominations; Jehovah's Witnesses criticize any images, including the cross; and the Amish forbid any image at all, including photographs.

> Let God come down and see we're only nine. He can count. And when he comes down, we'll count him in.
>
> RABBI OF A JEWISH CENTER IN THE BRONX, ON WAYS OF ACHIEVING A MINYAN, THE MINIMUM OF TEN MEN REQUIRED BY JEWISH LAW FOR WORSHIP

All of this, in addition to the issue of separation of church and state, is part of the controversy in the United States about the public display of tablets bearing the Ten Commandments. And even beyond: some religions number the Commandments differently or word them differently; if they're displayed in a public building, which version should be chosen?

The minyan is another of the powers of ten. In Hebrew, a minyan is a count, and it is the number necessary for public worship and communal prayer in a synagogue. In the Old Testament, ten was the number of righteous men that, had they existed, would have been sufficient to redeem Sodom and Gomorrah; ten, thus, can represent all Jews.

Orthodox Jews—the most conservative of the branches of the Jewish religion—believe that the minyan must consist of ten men; boys over thirteen, the age for Bar Mitzvah, are included as men. Some Conservative, Reform, and Reconstructionist temples count women as part of the minyan.

## PERFECT TENS

Given our fondness for our fingers, we often use their number to rank things—from one to ten, usually with ten as the highest possible score, though in some places the rankings go the other way; ten is the bottom and one is the best. Still, ten was the old Olympic high standard; it still ranks as "a perfect ten"—in the movie *10,* Bo Derek hit a ten on the scale of beauty and sex appeal.

Then there are the ten steps to this or that, the top ten whatever, the ten traits of enormously successful whoosy-whatsises, the first ten days, the big ten. . . . Football's Big Ten Conference is really eleven, but who wants to call it that? The Big Eleven just wouldn't sound the same. The seven-eleven fact is that the conference started when the presidents of seven midwestern universities met in Chicago in 1895 to discuss regulating and controlling intercollegiate athletics. Eventually—as one school left and others were added—it came to comprise Illinois, Indiana, Iowa, Michigan, Wisconsin, Michigan State, Minnesota, Northwestern, Ohio State, Penn State, and Purdue. I don't know what it says about the state of higher education when the Big Ten is actually eleven, but there it is.

Ten Downing Street is an address all its own—in Britain it

> When angry count to ten before you speak. If very angry, count to one hundred.
>
> THOMAS JEFFERSON

ranks with Parliament and Buckingham Palace, but its street address is much better known than either of those two. The prime minister lives there. Officially, it's the residence of the first lord of the treasury; it says so on the front door. The Downing Street house was given to Sir Robert Walpole on behalf of the nation and the Crown by George II. Walpole accepted, provided that the house was a gift to the incumbent first lord of the treasury rather than to him personally, and it remains the first lord's official residence. (Walpole moved in, after merging No. 10 with the house next door, in 1735.) The thing is that most prime ministers double as first lords of the treasury—in the twentieth century, the only prime minister who didn't was Lord Salisbury. Because of that, he didn't live at 10 Downing Street.

According to the Churchill Society of London, traces of both Roman and Saxon buildings have been found on the site of No. 10, and at one point in the Middle Ages, a brewery stood in its place. The last private resident of 10 Downing Street was a Mr. Chicken, "about whom little is known apart from his name."

The front door of No. 10 can be opened only from the inside—does the prime minister have to ring the bell? There's also an official country residence, Chequers, in addition to the London house—just as the president of the United States has Camp David in addition to the White House.

## COUNT TO TEN FIRST

Ten is a hardworking number. It may not be the perfect base for counting, not as adaptable as twelve or sixty, as magical as three, as pious as nine, or as mystical as eight—but it's ours. It's good and honest and true, simple and natural; it's the phallus and the egg.

Our ten fingers, after all, have served to count the universe. No matter what they're divided by, we love them, and they work. Given the state of what comes after our wrists, right there at the end of our arms, ten is our number, and it has served us well.

> Ten people who speak make more noise than ten thousand who are silent.
>
> **NAPOLEON BONAPARTE**

# 11

## ELEVEN IS A LEFTOVER NUMBER

We've used up all our fingers, and we still have one more to go. We need one toe, one cold toe, to reach eleven—the first number after our hands. It's also the first number in English with three syllables, a small fact but a fact nevertheless, and notable beyond that because it won't happen again until seventeen, six numbers away.

> **I don't feel I have to be loyal to one side or the other. I'm just asking questions.**
>
> **JUROR NUMBER 11,** *TWELVE ANGRY MEN*

Eleven has its own sense of oddness and mystery—partly because it's the first time we've left our hands, partly because it's the first of the double-digit numbers to be palindromic—it reads the same way backward and forward, and if you use just vertical lines for the ones, no little hook at the top or platform at the bottom, it's even the same upside down as it is right side up. There are other palindromic numbers—22, 33, 44, 55, 66, 77, 88, 99, 101, 111. . . . But eleven got there first.

The World Trade Towers were eleven incarnate, and an amazing eleven aura has risen up around them since the terrible day on which they fell. The cult of eleven begins with the attack that took place on September 11. In numbers, the date is known as 9/11, and nine plus one plus one is eleven. The first plane to hit the North Tower was American Airlines Flight 11. It had ninety-two passengers; nine plus two is eleven. Flight 77 hit the Pentagon; seventy-seven is eleven times seven; it held sixty-five passengers; five plus six is eleven. September 11 is the 254th day of the year; two plus five plus four is eleven. After September 11, 111 days remain until the end of the year. The emergency code for telephones is 911—nine-one-one, which is eleven, or nine-eleven. The state of New York was the eleventh state to join the Union; New York City is spelled with eleven letters. Afghanistan, where bin Laden was thought to be hiding, has eleven letters. The Twin Towers looked like an eleven. And each tower had 110 floors, which is eleven times ten.

Is this an amazing series of coincidences? Or, as some believe, is it a message from God? But surely, if God wanted to send us a message—about our wicked ways, or the coming of the Messiah, or the Second Coming and the imminent end of the world—he wouldn't send it in code. He spoke clearly—albeit through an angel—to Mary, among others; he spoke directly to Moses, and left his words chiseled in stone. Why would he bother with codes? Nearly three thousand people died on 9/11; many more suffered their loss and were left to grieve. It's hard to imagine a god counting the letters in New York City and deciding that those thousands of people, and all who loved them, should suffer as they did so that eleven could stand as a warning. Skywriting would have been a lot better.

In any case, the other two planes weren't elevens. United Airlines Flight 175 hit the South Tower; Flight 93 crashed in Pennsylvania. If God were sending us a message, the flight numbers would surely have matched.

World War I had something of the same mystery surrounding it. World War One is spelled with eleven letters. The War to End All Wars ended at 11 a.m. on November 11. The eleventh hour of the eleventh day of the eleventh month. The year, 1918, didn't match. It would have been better if the war had ended in 1911,

> This morning, the British Ambassador in Berlin handed the German Government a final note stating that, unless we heard from them by eleven o'clock that they were prepared at once to withdraw their troops from Poland, a state of war would exist between us.
>
> **NEVILLE CHAMBERLAIN,** PRIME MINISTER OF ENGLAND, IN A RADIO BROADCAST, SEPTEMBER 3, 1939

before it began. It was a terrible war; the losses were enormous; and it was certainly not the war that ended all wars. Armistice Day is still celebrated on November 11, but it's called Veterans Day in the United States and Remembrance Day in the Commonwealth of Nations and in many European countries.

## ONE MORE TO GO

The English word *eleven* is based on the idea of there being one more to count. At its simplest, it means one more than ten, one more than our fingers. It comes down from the Indo-European *leiqu* through the Greek *leipo* and the Latin *relinquere,* all variations of "to leave remaining," to the Anglo-Saxon *andelfene,* which breaks down into *andel,* or *ain,* or *one,* plus *lefene,* or *lef,* or *left.* One left. *Andelfene* goes to *endleofan* and then in pretty much a straight line to the Middle English *elleven,* whose two l's could *be* an eleven.

In Latin, *undecim* has the same concept of ten plus one, but it's not as obviously left over; it's simply one combined with ten, one-ten. The Romance languages, Italian, French, and Spanish, follow suit; in Italian, eleven is *undici;* in French, *onze* (*une* and *dix*); in Spanish, *once* (*uno* and *diez*). The German eleven, *elf,* like the English eleven, is related to *endleofan* or the Gothic *ain-lif* and the Old High German *einlif;* it certainly bears no resemblance to *zehn,* the German ten.

*Undecimal*—a handy English word to know if you have need of an unusual, elevenish sort of adjective—is obviously based on *undecim*. Undecimal counting would mean counting by elevens. There are people who have eleven fingers, though they are few and far between, and for them undecimal counting would come more or less naturally. *Undecennary* is a word that doubles as a noun and an adjective; as a noun it means a period of eleven years, or an eleventh anniversary. Not too far away, as an adjective, it means occurring every eleven years, or pertaining to eleven years. It combines the Latin *undecim* with *annus,* "year."

In Greek, eleven was *hendeca*—from *hen,* the neuter of *heis,* "one." From *hendeca* comes *hendecasyllabic,* a mouthful of a word that means just what it says: having eleven syllables. "A word or line of eleven syllables" is a phrase that contains exactly eleven syllables—it's hendecasyllabic.

Ten was once the limit of counting, the end of our fingers. Anything beyond that was—here the music swells—the grand and elusive *many,* which was pushed higher and higher as we used more and more numbers. One, two, *many;* one, two, three, *many;* one, two, three, four, *many;* and now, one to ten and then *many.* We continued with counting's inevitable march forward after we had stopped to breathe at ten because we still (as we always do) had one more, one left, to count. *Endleofan.* One left.

That doesn't make eleven a very distinguished number. Derivatively, at least, it's just a bowl of leftovers. It does have a certain kind of mathematical solidity. Eleven is a prime number. Prime numbers aren't rare, though—two, three, five, and seven are all prime numbers; they can only be divided evenly by themselves and one. To divide evenly means there's nothing left over, and eleven is very much a leftover number.

But eleven is a kind of bridge—between the limits of our fingers and the universe that lies beyond. There was a time (and it hasn't really passed yet) when it seemed that adding one to an already large number might turn it into an enormous quantity. That's what eleven is. Ten and one more. Forever and a day.

Eleven is, of course, a pair. It's parallel lines—length-challenged, to be sure. It's a male-bonded number: a pair of pricks. Maybe that's why there are eleven men on football, soccer, rugby, and cricket teams. There isn't anything soft about eleven. Eleven on the playing fields of Eton. Or off.

## COME ELEVEN

Even aside from ball games, there are more clusters of eleven than you might think. Eleven states seceded from the Union to become the Confederate States of America and fight the Civil War: Alabama, Arkansas, Florida, Georgia, Louisiana, Mississippi, North Carolina, South Carolina, Tennessee, Texas, and Virginia.

A Norwegian musical spirit named Strömkarl had eleven different tunes in his repertoire. Anyone could—and did—dance to ten of the melodies, but the eleventh belonged to the night spirit. If it were played, forced to dance would be tables and benches, cups and cans, old men and old women, babies still in their cradles, the blind and the deaf, and the sick in their beds.

According to the Peripatetics, followers of Aristotle, the universe consisted of eleven spheres, each nestled into the one before it, like Chinese boxes or Russian dolls. The final two spheres—in addition to those mentioned in the chapter on nine (the sun, the moon, Mercury, Venus, Mars, Jupiter, Saturn, the firmament, where the stars are, and the crystalline sphere)—are the primum mobile and the empyrean, the highest heaven, heaven of the blessed. The primum mobile (Latin for "first moved") was believed to revolve around the earth in a different direction from the other spheres. The empyrean was paradise, the sphere of the blessed—a happy eleven.

The Fibonacci numbers have a happy—and odd—eleven quirk. Any ten Fibonacci numbers in a sequence will always add up to a number that's divisible by eleven. The first ten numbers:

1+1+2+3+5+8+13+21+34+55=143, and 143 divided by 11 is 13. Or, to use higher numbers, continuing the sequence: 55+89+144+233+377+610+987+1,597+2,584+4,181= 10,857, and 10,857 divided by 11 is 987. There's more: the sum of any sequence of ten Fibonacci numbers when divided by eleven always gives a number that is the seventh number in the sequence. In the first example above, 13 is the seventh number in the sequence; in the second, 987 is. Is this more Fibonacci magic? Absolutely.

Less pleasant is the view of eleven in the play *Die Piccolomini*, by the German poet Friedrich Schiller. In it, an astrologer declares, "Eleven is the sin. Eleven transgresses the Ten Commandments." Two is the sin, the step away from God, which is one; eleven is the step away from the holiness of the Pythagorean ten; and now, eleven is also the sin one step away from the Ten Commandments. Poor eleven! Less round and useful than the sinful two—not as well known, perhaps, but out there trying its best, and still being called a sin. What about all the other numbers, just that much farther away, each in its turn? Are they *all* sinners, compounding the wickedness, once we've passed ten? Surely not.

## ELEVENSES

*Apollo 11* was the first—and is so far the only—manned spacecraft to land on the moon. On July 20, 1969, mission commander Neil Armstrong and lunar module pilot Buzz Aldrin took those unforgettable first steps on the moon, while command module pilot Michael Collins orbited, waiting to bring them back home. For fans of eleven: sixty-six (eleven times six) years earlier in 1903, the Wright Brothers flew the first powered aircraft—just a few feet off the ground in Kitty Hawk, North Carolina.

There are other singular elevens: There are eleven stars in

> Gentlemen, Chicolini here may talk like an idiot, and look like an idiot, but don't let that fool you: he really is an idiot. I implore you, send him back to his father and brothers, who are waiting for him with open arms in the penitentiary. I suggest that we give him ten years in Leavenworth, or eleven years in Twelveworth.
>
> **GROUCHO MARX,** AS RUFUS T.FIREFLY, *DUCK SOUP,* 1933

Vincent van Gogh's painting *Starry Night.* One of the two classic palindromes, Madam I'm Adam, has eleven letters. (The other, Able was I ere I saw Elba, has nineteen, a number of little consequence here.) The first 7-Eleven store was open from seven in the morning until eleven at night. The secret formula for Kentucky Fried Chicken is supposed to have eleven ingredients—I assume that doesn't count the chicken, which ought not to be a secret. The language Rotokas has only eleven letters; it is spoken on the South Pacific island of Bougainville. The win-or-lose toss in a game of craps is eleven. The cycle of sunspots is undecennary—which you'll remember means based on eleven years.

By now, we could be hungry for elevenses—a fine time for a midmorning cup of tea and a snack in the United Kingdom and most of the Commonwealth countries. Elevenses were beloved of Winnie the Pooh, who liked his tea and honey at any time of day, when he felt elevenish. (Perhaps elevenses are a double plural of eleven, an after-breakfast between-meals moment.) Paddington Bear was also fond of elevenses—it's not so much that bears like their snacks (though indeed they do); it's more in the nature of children's books. In *The Lord of the Rings,* Merry and Pip are not happy to discover that they might miss elevenses on their way to help Frodo cast the One Ring into the fires of Mount Doom.

Bilbo Baggins celebrates his birthday at the beginning of *The Lord of the Rings.* "Bilbo was going to be eleventy-one, a rather curious number, and a very respectable age for a Hobbit," wrote

J. R. R. Tolkien. Or as Bilbo puts it, "Today is my one hundred and eleventh birthday. I am eleventy-one today." (For the record, eleventy-one has exactly eleven letters. Please don't count the hyphen; it's not a letter.) Lebenty-leben, only slightly different, is from Uncle Remus, and it's a children's word—as when counting: one, three, ten, lebenty-leben. (For that matter, why isn't eleven pronounced onety-one?)

> **Eleven syllables, many of them of Latin or Greek derivation, when one good English word, a Saxon word of a single syllable, would do.**
>
> JOSEPH CHAMBERLAIN, COMMENTING ON CHURCHILL'S PHRASE "TERMINOLOGICAL INEXACTITUDE," TO REFER TO THE WORD "SLAVERY," FEBRUARY 22, 1906

There's a kind of poem that contains exactly eleven words. It's called an *Elfchen* in German, which means "little eleven." It has one word on the first line, two words on the second line, three words on the third, four words on the fourth—and one word on the fifth line. As in:

*Elf*
*means eleven*
*in German but*
*in English, we think*
*elves.*

And now, at the eleventh hour, there is the eleventh hour—the hour that is just in time, or at the last possible moment, from a parable in the Bible involving laborers hired at the eleventh hour of a twelve-hour workday, yet paid the same amount as those who had worked the whole day. Because eleven is the hour before midnight, the eleventh hour can also imply urgency, or an impending emergency.

> **And I heard a Bugle sounding, as from some celestial Tower; And the same mysterious voice said: "It is the Eleventh Hour!"**
>
> FORCEYTHE WILLSON, "THE OLD SERGEANT"

There isn't much more to be said about eleven. For all of that, from elevenses to the eleventh hour, it remains a number without the kind of dignity and distinction that belong to most of its predecessors. Still, what it has matters enormously. It's more than ten—that number of perfection—and while it doesn't have the roundness and usefulness of twelve, it has its own place in the order of things. Perhaps what it does best is to get us there. It makes the leap past ten; it carries us through; it fills the gap; it does the thing. Eleven is a number of necessity. We could not get to twelve without it. It is indeed the father—doubly phallic—of twelve. And that has both dignity and distinction. We need eleven, and that's all there is to it.

# 12

## TWELVE IS THE LAST OF THE BASIC NUMBER NAMES

There's a lovely anagram about eleven and twelve: Eleven plus two equals twelve plus one. True mathematically—both equal thirteen—but here, "equals" also means that if you scramble up the letters on one side of that word, you can reshuffle them and come up with the letters on the other side. You can make *twelveone* out of *eleventwo.* Why would you want to do that? Just for fun. After all, they're both leftover numbers, linguistically and mathematically. After *endleofan* came *twelf:* a combination of *twa,* for two, and *lif,* for left. Two left after you've used up your ten fingers.

> My living in Yorkshire was so far out of the way that it was actually twelve miles from a lemon.
>
> **SYDNEY SMITH,** *LADY HOLLAND'S MEMOIR*

But that isn't accurate. You *can* count to twelve on your fingers—many people did, and some still do. Instead of thinking of your fingers as whole units, with five verticals on each hand, regard them as bendable and then count the bends. Each finger (except the thumb) has three bends—three phalanxes joined by the knuckles—for a total of twelve spaces that can easily be counted, with the thumb used as a movable pointer. Going higher is easy, too, in the same way that it's easy to count multiples of ten after we've used all our fingers. There are dozens of alternatives: one either remembers, or makes a note, or moves a bead, or (my favorite) uses one hand for counting twelves and the other for keeping track—counting one to twelve on the right hand, and then folding down a finger on the left to indicate a total of twelve. Another twelve on the right hand equals two folded fingers on the left, and so on up to five folded fingers, for a total of sixty—which may go some distance toward explaining base sixty counts.

A dozen is twelve, and it's one of the many leftover ways that show we once counted by twelves. Eggs come by the dozen, and so do oysters and a variety of other things that we never stop to think about. A gross is a dozen dozens. A ruler has twelve inches. A year has twelve months. The clock is divided into twelve hours, a day into twenty-four, two sets of twelve.

England was probably the most serious about counting by twelves—until 1971, the British monetary system was twelve-based. The pound sterling was, and is, the basic currency of the UK. The difference between before what's called Decimal Day and now is that now the pound is divided into units of tens, and before it was divided into twelves. For a while after the change, parents lamented losing the ease with which their children had learned fractions: they had once learned twelfths, sixths, fourths, thirds, eighteenths, halves, and a bunch of other fractions as naturally as American children learn quarters and dimes—which are really fourths and tenths.

> Annual income twenty pounds, annual expenditure nineteen nineteen six, result happiness. Annual income twenty pounds, annual expenditure twenty pounds ought and six, result misery.
>
> **CHARLES DICKENS,** *DAVID COPPERFIELD*

The English pound in the duodecimal system (adjective, relating to the number twelve, from the Latin *duodecimus,* "twelfth," from *duodecim,* "twelve," from *duo,* "two," and *decem,* "ten") was divided into 240 pence. The singular of pence is penny, and the symbol for penny is p, so for the five pence, the Brits write "5 p," which they pronounce "five pee."

Each twelve pence was equal to one shilling, and twenty shillings made a pound. Which doesn't say anything about farthings, bobs, florins, crowns, or guineas. There were four farthings to a penny (or two halfpennies). Farthings, halfpennies, and pennies were all known as coppers—because, yes, they were all made of copper. A florin was worth two shillings. A shilling was the same as a bob (and bob wasn't used in the plural—fifteen shillings was fifteen bob); a five-shilling piece was a crown; a guinea was worth just over a pound. A sovereign was a gold pound coin. *Quid* is the slang term for pound; the plural of pound is pounds, but the plural of quid is still quid. Confused? If not, we shall add that sometimes British money is called sterling, which is short for pounds sterling, from the days when the pound was equal in value to one pound (in weight) of silver. Those days are long gone.

> **Wery glad to see you, indeed, and hope our acquaintance may be a long 'un, as the gen'l'm'n said to the fi' pun' note.**
>
> **CHARLES DICKENS,** SAM WELLER, IN *THE PICKWICK PAPERS*

The symbol for the British pound is £—a fancy sort of L with a line through its middle. The word *pound* is an English translation of the Latin *libra.* Shillings are abbreviated by an s, not from shilling, as you'd expect, but from the Latin *solidus,* a Roman coin, and pennies are represented by d, from another Roman coin, the *denarius.*

## WORDS

There are more words than one would expect based on twelve—like duodecimal, already used about the old British currency system. The *duodenum* is the first portion of the small intestine; its length is approximately equal to a twelve-finger breadth; in German, it's called the *Zwölffingerdarm,* the "twelve-finger intestine." *Duodecennial* is a handy word for a twelfth anniversary, in case you don't want to say twelfth anniversary. When it's used as an adjective, *duodecennial* means "of or pertaining to twelve years." *Duodecimal,* when it isn't being used as an adjective—of or relating to the number twelve—is a noun that means a twelfth.

*Unciary* also means a twelfth part—from the Latin *unciarius,* in turn from *uncia,* a twelfth part—and it's the source of the words *ounce* and *inch*. The original pound was the troy pound, which had twelve ounces. It's probably named for the French market town of Troyes, where English merchants traded as far back as the early ninth century. Troy ounces are now used only for weighing precious metals like gold, silver, and platinum, and for certain jewel stones, like opals. Twelve troy ounces equal 240 penny-weight—the base of the old British penny—the weight of a sterling silver penny, as adopted by King Henry II in the twelfth century. One sterling silver penny weighed 1/240 of a troy pound, and that's full circle.

An inch—also from *unciary*—has a tangled derivative history. It could be the distance between the tip and the first joint of the thumb—not exact, perhaps, but close. In several languages, the word for inch is close to the word for thumb: in French, *pouce* is inch as well as thumb; in Italian *pollice* is both; in Dutch, *duim* is both; in Spanish, *pulgada* is inch, and *pulgar* is thumb; in Portuguese, *polegada* is inch and *polegar* is thumb; in Swedish, *tum* is inch; *tumme,* thumb; in Sanskrit, *angulam* is inch, and *anguli* is finger. The thumb is indeed a handy measure—as is the human foot and the arm, all once used as measurements, even though your foot may not be the same size as mine. In *World Wide Words,* Michael Quinion notes that in medieval England, an act of Parliament defined an inch to be the breadth of the thumb belonging to the official alnager, a man who was responsible for measuring and stamping pieces of cloth, thus protecting buyers of woolen goods (an important English export) from fraud. The title "alnager" takes its name from the Old French *aulne,* to measure by the ell—and ell is a word that derives from the same Indo-European root word as does *ulna,* Latin for "forearm." And an ell, it should also be noted, meant "arm" in Old English, which is why we bend our arms at the "elbow."

*Nychthemeron* is a word which has no direct relationship to twelve's linguistic ancestors, but it relates to twelve nonetheless. It means a full period of twenty-four hours, a night and a day of twelve hours each, and it comes from the Greek words *nyktos,* "night," and *hemera,* "day." (Night comes first in *nychthemeron* because the Greeks believed the day began at sunset.) *Nychthemeron* is related to *circadian* (from the Latin *circa,* "about," and *dies,* "a day"), but circadian refers to the body's clock, while nychthemeral rhythms are based on the world outside the body. Circadian cycles are largely genetic, but they're affected by daylight as well as by artificial light. When we have jet lag—or when the clock is moved forward or back an hour or if we're at our computers too long—our circadian and nychthemeral rhythms are jarred against each other, and that does not feel good.

In its own category is *dozen*—a nice, simple word we all know, still based on twelve. Through French, *dozen* traces back to the Latin *duodecim*: the French word for dozen is *douzaine,* and that's a lot closer to a carton of eggs than that long Latin word. *Douzaine* means "about" twelve, rather than exactly (twelve is *douze*)—but still; close enough. *Vingtaine* means about twenty (twenty in French is *vingt*); *centaine*

> **What vexes me most is, that my female friends, who could bear me very well a dozen years ago, have now forsaken me, although I am not so old in proportion to them as I formerly was: which I can prove by arithmetic, for then I was double their age, which now I am not.**
>
> JONATHAN SWIFT, IN A LETTER TO ALEXANDER POPE, FEBRUARY 7, 1736

means about a hundred (cent). . . . They're handy words, in their way, summing up an idea in a simple couple of syllables.

Twelve dozen is a gross (numbering 144); twelve gross (1,728) is called a great gross (not grossly gross). A great hundred is 120, or ten dozen (twelve on each finger on both hands).

A baker's dozen, on the other hand, is thirteen. In thirteenth-century England, bakers who were found to be shortchanging their customers could be given severe punishments. Just to be extra careful, bakers began giving thirteen for the price of twelve. At the same time baking thirteen for every twelve meant the baker himself was protected against burned or lost rolls, or the possibility that a hungry apprentice might eat one. Today, thirteen cookies or biscuits fit a baking pan as easily as twelve do, but since the corners of the baking pan get hotter than the rest, the best way to bake is in rows of four-five-four. That makes thirteen. "Wiley's Dictionary," in the comic strip *B.C.,* defined a baker's dozen as "Twelve of today's and one of yesterday's." There's that, too.

## TEENS AND -TYS

In English, there are very few number names that stand alone, uncomplicated by added syllables. They are one, two, three, four, five, six, seven, eight, nine, ten, eleven, and twelve. (And hundred, thousand, million, and googol, but we haven't reached those yet.) Between twelve and hundred, every single number name is a combination of the numbers that have gone before. (The same is true for the numbers between hundred and thousand, and then thousand and million.) Our remarkable system uses very few numerals to count all of our numbers.

The English connectors—the syllables that make new numbers out of old ones—are only two: -teen and -ty; "-teen" is an old form of ten, so all the teens are ten plus a number. Thirteen is three plus ten (it was first pronounced as "threeteen" and written as "treteen"); fourteen is four plus ten, and so on through nineteen. We could instead have continued with leftover names, as we did with eleven and twelve—but having gotten that far, we seem to have realized there was a lot further to go, and leftover wasn't going to be sufficient to explain what we were doing. There are languages that begin the compounding after ten—for them, *many* began with the number after ten. For us, the numbers past twelve may once have been seen in that way, like the old classifications of one, two, and *many*—singular, dual, and plural. One-plus-ten and two-plus-ten were not yet in the category of *many*—but thirteen was, just as three was.

German and Dutch count the teens the way that English does. In German, eleven is *elf* (through the Gothic *ain-lif,* "one left," and the Old High German *ein-lif*); twelve is *zwölf* (through *twalif* and *zwelif*), but thirteen is *dreizehn* (*drei* is three and *zehn* is ten), fourteen is *vierzehn,* fifteen is *fünfzehn,* and so on through *sechszehn, siebzehn, achtzehn,* and *neunzehn*. The rule holds true for the Germanic languages.

The Romance languages do it differently. They don't mark leftover numbers; they reach the concept of *many* right after ten. Eleven is French is *onze,* in Spanish it's *once,* in Italian, *undici* (all from the Latin, *undecim,* all thus really meaning one-and-ten). Twelve in the same order of languages is *douze, doce,* and *duodici*. Thirteen: *treize, trece,* and *tredici*. Fourteen: *quatorze, catorce,* and *quattordice*. After sixteen, the "ten" syllable is clear: at seventeen, French is *dix-sept* and Italian is *diece-sette*—and so on through nineteen.

At twenty, we leave the teens behind and find instead another derivative of ten: -ty, which combines with two for twenty, three for thirty. . . . German uses *-zig,* or *-sig,* as in *zwanzig* for twenty, *dreissig* for thirty. . . . The Romance languages transform the Latin suffix *ginta,* a transforming ten, into just *-a* or *-e,* as in the French *trente* for thirty, the Italian *trenta,* and the Spanish *treinta.* (French for twenty is *vingt,* based on the Latin *viginti;* it's *venti* in Italian and *veinte* in Spanish.) French becomes irregular at seventy, which is based on sixty-plus-ten, *soixante-dix;* and it's based on twenty at eighty, *quatre-vingts,* four-twenties. Ninety is *quatre-vingt-dix,* four-twenty-plus-ten.

This is what it comes down to: All the numbers up to ten have their own names. In a few languages, the next two numbers have their own names too. But generally, after ten, only three multiples of ten (100, 1,000, 1,000,000) have their own names. All the other number names are combinations of the names that have gone before. It's a kind of verbal-mathematical magic.

## KEEPING TIME

Among our most important—certainly our most useful—twelves are the ones by which we measure our lives, the moments, days, and months of our years. Over the centuries we've marked time in several ways, from the smallest hourglass to the monoliths at Stonehenge. Most often, our sense of time is based on what we see in the sky—the movement of the stars, the sun, and the moon—but not always. In ancient Egypt, the year was divided into thirds: the flooding of the Nile, the time for seeding, and the time for harvest. In tropical countries, the rainy season and the dry season often delineated the passage of time. There are places where the cry of migrating cranes meant it was time to plow and to sow. In another spot, when flying fish appeared, the fishing time of year began. Keeping track of the tides is another way to measure time. But most often, it's the seasons of the moon, as it waxes and wanes, shines and disappears, that are

> Twelve significant photographs in any one year is a good crop.
>
> ANSEL ADAMS

the basis for measuring time. There are some tally sticks that researchers believe meant the counting of days from moon to moon was happening as early as the Paleolithic age. The English word *month* is related to the word *moon,* and both come from a third word, meaning "to measure."

A year of time can be measured not only by the full moon, which takes about a month from round circle to round circle, but also by the sun, because it takes the earth a year to travel around the sun, and by the phases of the moon, each of which lasts about seven days. There's a problem with the three systems: they don't overlap, and they aren't exact. There aren't precisely twelve full moons in a year, or exactly seven days in each phase of the moon. (Full moon to full moon is about twenty-nine and a half days.) That's why we have leap years—to adjust our calendars every four years—and it's also why people in other places use different calendars, measuring time in different ways.

> **Love, all alike, no**
> **seasons knows,**
> **nor clime,**
> **Nor hours, days,**
> **months, which are**
> **the rags of time.**
>
> **JOHN DONNE**

Our year is solar, the time it takes the earth to revolve around the sun, and it combines the orbit of the moon with seasons based on the sun. The months are related to the moon—but not *exactly*—and the year is related to the sun. The months alternate days—thirty-one and thirty—except for July and August, which both have thirty-one, and February, which was shortchanged to keep things even at the end of the year, except for leap year, when it's lengthened to keep the seasons consistent over the centuries.

Children learn "Thirty days hath September . . ." to help them remember which months have thirty days and which have thirty-one, but there's an even easier system, a kind of finger counting. Using the knuckles (or the fingers) on one hand as well as the spaces between the fingers, begin with the knuckles of your index finger. That's January, and it has thirty-one days; next is

February, and you'll just have to remember whether it's leap year or not. The next finger is March, for thirty-one days; the next space is April for thirty; then May for thirty-one; followed by the space for June, thirty. Last finger is the pinkie, which is July, for thirty-one days. Now go back to your first knuckle and this time it's August, and it's thirty-one days again. Knuckle months have thirty-one days; space months, except for February, have thirty.

Our calendar is based on the Roman calendar (the Latin word is *calends*), which originally began on March 1. March 1 is really a fine time for New Year's Day—spring, in the form of the vernal equinox in the Northern Hemisphere, arrives in March, and how much more sensible it would be to celebrate a new year in springtime than in the dead of winter, as we do now.

> Time has no divisions to mark its passage, there is never a thunderstorm or blare of trumpets to announce the beginning of a new month or year. Even when a new century begins it is only we mortals who ring bells and fire off pistols.
>
> THOMAS MANN

If you've been reading carefully, you know September, October, November, and December are numerically illiterate names. September isn't the seventh month; October isn't eighth; November isn't ninth; and December isn't tenth. How did we get so sloppy? You could blame it on the Romans—but you'd be wrong if you did, because they began by doing it right. The Roman calendar we know best was introduced by Julius Caesar in 46 BC—though in legend it was invented by Romulus, the founder of Rome in about 753 BC, and in fact was probably based on the Greek lunar calendar. When the Roman calendar began in March, September was the seventh month and December the tenth, just as their names tell us. It was only later, by centuries, that January and February made it to the number one and two spots and the other months had to move out of the way.

January takes its name from the two-faced Roman god Janus,

who opened the gates of heaven at dawn to let out the morning, and closed them again at dusk. He could look backward and forward at the same time, a good quality for bringing in the new year. Most of the Roman months were named after gods. But there was a hitch—related, in a way, to the early counting of one-two-three-four-*many*. (The Latin words for one, two, three, and four are declined, changing forms according to their use in a sentence. But after four, the number word stayed the same no matter how it was used.) In a Roman family, the first four children were given individual names—like Julius, Marcus, Cassius. . . . After the first four, naming the fifth child was easy: Quintus was the fifth boy; Sextus was the sixth; Septimus was the seventh. (Some names like those are still used. Onzieme is a French name for an eleventh child. Balinese children are named in cycles of four—not for numbers, but rather to indicate an imperishable form: human beings may come and go, but their traditions are meant to last forever.)

> **What's good about March? Well, for one thing, it keeps February and April apart.**
>
> **WALT KELLY**

July and August were originally called Quintilis and Sextilis—yes, because they were the fifth and sixth months. Their names were changed to honor Rome's great emperors, July for Julius Caesar and August for Augustus Caesar.

In the western calendar, in which the months hold hands with the moon, most months have only one moon to cherish. But once in a while, because the months don't exactly match the cycles of the moon, the days back up and there are two full moons in the same month. It happens every so often—not rarely, but not frequently—and when it does, we call the second moon a blue moon. There's a second explanation for the phrase about blue moons (once in a . . . ), and that is that there are three months in each season, and usually the same number of full moons—one full moon to each month. But once in a while, there are four moons in a single season, and the fourth is called the blue moon. Either way, blue moons don't happen very often.

Aside from the original Julian and the later Gregorian calendars, there have been—and still are—others along the way. The French Republic had a calendar of its of its own, used for about twelve years after it was adopted in 1793. It had twelve months; each month had thirty days; they were arranged in three ten-day weeks called *décades,* and extra days were added at the end of each year to take care of the gaps. The year began with the autumn group, but the spring *décade* had particularly lovely names: Germinal, Floréal, and Prairial. Thermidor and Fructidor were two of the three summer months.

The Hebrew calendar is based in part on the cycles of the moon, and the Islamic calendar is truly lunar, which is why the months and the holy days in each month are at different times each year. Months in Hindu calendars (also based on both lunar and solar years, as is the Chinese calendar) are sometimes named for the zodiac sign in which the sun is traveling. The Persian calendar, used in Iran and Afghanistan, also has twelve months. The old Icelandic calendar isn't used anymore, but some holidays are still based on its count. It had twelve months, in two groups of six each, called Short Days and Nightless Days. The Short Days were from mid-October through mid-April. Mid-December to mid-January was Yule Month; next was Fat-Sucker Month, a hungry time of year. Innards Month was for slaughtering of livestock, when everything that could not be pickled or salted for the winter needed to be eaten. The Nightless Days of summer included Sun Month, when the midnight sun stayed above the horizon, and Hay Working Month, from mid-July to mid-August.

We've read about the seven days of our week in the chapter about seven. But our smallest units of time—the seconds, minutes, and hours—are measured differently from the days, months, and years: by clocks (from the Latin *clocca,* "bell," from when bells tolled the hours) that count the hours over a twelve- or twenty-four-hour day.

Sundials were early timekeepers, marking off the hours by the shadows cast by the sun. Candles or incense that burn down at predictable speeds, sand that pours through a tiny hole in an hourglass, water that flows at a constant rate—all have been used to measure the passing moments.

Some early tower clocks didn't use minute hands; instead, they told the time with audible signals, like bells. Thirteenth-century Italian monks used tower clocks to tell the hours of prayer. Some sixteenth-century clocks, still telling time today, show only the quarter hours, telling time to the nearest fifteen minutes. Galileo thought of the first pendulum clock in the seventeenth century; grandfather clocks were invented later to house the pendulum and its works, and soon after that, a ticking clock was patented.

A full day of twenty-four hours marks the turning of the earth on its axis. For most of us, the day begins with dawn and ends at sunset—but not for all and not for always. In ancient Egypt, a day was counted from sunrise to sunrise. The Jewish day begins at sunset, and medieval Europe followed this same tradition. Christmas Eve, New Year's Eve, and Halloween are remnants of the time when holidays began in the evening, rather than in the morning.

Muslims fast from dawn to sunset of each day during the month of Ramadan, and traditionally count dawn as the moment when a white thread can be distinguished from a black thread, using only natural light.

## HALLOWED TWELVES

Twelve is a good number—even for those of us with only ten fingers. It has a sense of roundness and completeness, just as ten does, and not only because it can be divided up evenly into so

> Are twelve wise men more wise than one? or will twelve fools, put together, make one sage? Are twelve honest men more honest than one?
>
> **HERMAN MELVILLE,** *MARDI*

many nice chunks. As the union of one and two, it's both male and female, both odd and even. That doesn't make it the first hermaphroditic number; on the contrary, it's very nicely sexed, thank you, and has had a long and happy marriage.

*Twelve* is also an absolute symphony of consonants. It's a word that's satisfying to say slowly, savoring the *tw* and *lv* as they roll over your tongue and lips. To a child it might have the magical sound of "two elves." *Twelve* has the feeling and sound of an ancient word, and indeed, it has an old and honorable pedigree.

Most obviously, one thinks of the twelve disciples of Jesus, first identified as "the twelve" in the Gospel of Mark. They were Peter, James (son of Zebedee), John, Andrew, Philip, Bartholomew, Matthew, Thomas, James (son of Alphaeus), Thaddeus (or Judas not Iscariot), Simon (the Cananaean, or the Zealot), and Judas Iscariot. Traditionally, it was Judas Iscariot who betrayed Jesus, but the recently discovered Gospel of Judas says instead that Iscariot was deliberately chosen by Jesus as the one for the necessary act of betrayal—that he was, in fact, the disciple closest to Jesus. After the act of betrayal set forth in the Gospels, Judas Iscariot was replaced by Matthias. The disciples were later called the apostles, from a Greek word, *apostolos,* that means "the person sent forth." The pope, as head of the Catholic Church, addresses the prelates of the church as "the twelve," though they number in the hundreds.

> Among twelve apostles there must always be one who is as hard as stone, so that the new church may be built upon him.
>
> **FRIEDRICH NIETZSCHE,** *THE WANDERER AND HIS SHADOW,* APHORISM 76, "THE MOST NECESSARY APOSTLE"

> At Christmas I no
> more desire a rose
> Than wish a snow in
> May's new-fangled
> mirth;
> But like of each thing
> that in season
> grows.
>
> **WILLIAM SHAKESPEARE,** *LOVE'S LABOUR'S LOST*

Less obviously, from the point of view of this hallowed number, there are the twelve days of Christmas, celebrated in the most quotidian way by the seemingly endless verses of the song which begins and ends with that poor partridge, stuck forever in the pear tree's branches. Of course, at Christmas, unless you're so southerly that it doesn't matter in any case, a pear tree has no leaves and certainly no pears. But that's the least of the dilemmas posed by the days and their weird collections of gifts. Would you really like to be given eleven pipers piping?

There are those who maintain that the song is actually a hidden text of instruction with roots in the sixteenth-century religious wars in Britain, when it was a crime to be Catholic. From 1558 to 1829, being caught with anything written that showed belief in Catholicism meant possible imprisonment, hanging, or being drawn and quartered—a literal phrase that meant a death of the most gruesome sort.

According to this theory, the song was a mnemonic device to teach children the catechism. The "true love" who gives all the gifts is God, and the "me" who receives them is each person baptized in the faith. The partridge stands for Jesus; the two turtle doves are the Old and New Testaments; the three French hens are the virtues of faith, hope, and charity. The four calling birds stand for the four Gospels or the Four Evangelists; the five golden rings are the first books of the Old Testament, the Pentateuch, or the Torah—Genesis, Exodus, Leviticus, Numbers, and Deuteronomy—together, the story of man's fall from grace. The six geese are the days of creation; the seven swans, the gifts of the Holy Spirit, or the seven sacraments; the eight maids, the beatitudes. Nine ladies stand for the fruits of the Holy Spirit; ten lords are the Ten Commandments; eleven pipers, the faithful apostles; and twelve drummers, the points of doctrine in the Apostles' Creed.

Can it be so? Well, there are differences of opinion. Most obviously, there is nothing in the song that is exclusive to Catholics. The things that are (confession, for example, and allegiance to the pope) don't surface in the song. Then, it's sung only at Christmas—how would children learn and remember during the rest of the year? The relationship between the numbers and what they might represent is somewhat arbitrary—and is not always explained in the same way.

Historically, the song first saw print as a memory game in a 1780 children's book, *Mirth Without Mischief.* The leader was to sing the chorus line—on the (blank) day of Christmas—and the players were to sing the rest. The first person to make a mistake was to forfeit something—a sweet or, better, a kiss.

The song existed before the publication of the book; there's a possibility it was originally French. Partridges were first brought to England by the French in about 1770. On another level, "calling birds" are thought originally to have been colly, or collie, birds—colly meant as black as coal (like collier, a coal miner, or colliery, a mine), so colly birds would have been blackbirds. By the same token, golden rings may originally have meant not jewelry, but five ring-necked birds like pheasants—and in that case, the first seven gifts (the partridge, turtle doves, French hens, calling birds, golden rings, geese, and swans) were all birds, a happy conceit.

The twelve days of Christmas begin for most Christians with December 26 and end on January 6, Epiphany, marking the visit of the Magi—the Three Wise Men, Caspar, Melchior, and Balthasar, with their gifts for the newborn child, thus showing him to the world as Lord and King. In Hispanic and Latin cultures, and in some parts of Europe, Epiphany is known as Three Kings' Day. In Eastern Orthodox churches, Epiphany emphasizes the baptism of

Jesus in the Jordan River by Saint John the Baptist, marking the only time when all of the Holy Trinity manifested their presence to humanity simultaneously—God the Father spoke through the clouds as God the Son was baptized in the river, and God the Holy Spirit flew over the scene in the shape of a dove. Orthodox churches have a Blessing of the Waters ceremony on Epiphany, when the priest throws a cross into the nearest body of water, and whoever swims out, finds it, and brings it back receives a special blessing and is said to be going to have good luck throughout the coming year.

> **I know the answer!**
> **The answer lies**
> **within the heart of**
> **all mankind!**
> **The answer is twelve?**
> **I think I'm in the**
> **wrong building.**
>
> CHARLES M. SCHULZ
> CREATOR OF *PEANUTS*

In some places, gifts are given on each of the twelve days of Christmas, just as the song tells us. In some churches, Christmas decorations are traditionally removed on Twelfth Night. There are a host of Epiphany traditions in various countries—in France and Belgium, there's the *galette des Rois,* in Mexico, the *Rosca de Reyes*—all cakes or sweet breads with a toy or bean baked inside; whoever has the piece with the treasure in it is king for a day in France or Belgium, and in Mexico, has to host a party about a month later, on Candelaria Day.

## AND STILL TWELVES

The knights of King Arthur's round table—a real table as well as the highest order of chivalry at the king's court—numbered way more than twelve, but the ones closest to the king's heart were traditionally numbered at an even dozen. The table itself was in Camelot, and it was deliberately designed as a circle to avoid pride of place and resultant jealousy among the knights. It was big enough—big!—to seat a hundred and fifty knights. No matter where they sat at the table, twelve knights held the very highest honor. Among the twelve knights of the round table were such as Lancelot, Tristram, Galahad, Gawain, Kay, and Percival.

> Urge no healths. Profane no divine ordinances. Touch no state matters. Reveal no secrets. Pick no quarrels. Make no comparisons. Maintain no ill opinions. Keep no bad company. Encourage no vice. Make no long meals. Repeat no grievances. Lay no wagers.
>
> THE TWELVE GOOD RULES OF KING CHARLES I

The twelve steps of Alcoholics Anonymous and AlAnon are another dozen. Bill W. and Dr. Bob S. cofounded AA in Akron, Ohio, in 1935; today this voluntary fellowship has over two million members who share their strength, experience, and hope in more than one hundred thousand AA groups in 150 countries around the world. The AA steps are both practical and spiritual, but they are not religious. Their reference to a higher power can be read as God, or as any force outside of oneself—nature, humanism, even the group itself. The first step makes the most basic statement of the twelve: "We admitted we are powerless over alcohol, and that our lives had become unmanageable." In that step is both an admission of powerlessness and the dawning realization of enormous strength—the power to change oneself.

American schools are divided into twelve grades (not counting kindergarten, which is not a legal requirement). Children begin first grade when they're six and emerge twelve years later from high school when they're eighteen; in educational jargon, it's a K-12 system. It evolved from private academies, Latin grammar schools, reading and writing schools, and home schools. Public education and a free press, said Jefferson, are the necessities of a democracy, and they are both pillars of freedom.

The twelve-person jury system as we know it is another of freedom's pillars. There are references to jury trials in the Bible. In one of the most famous trials in history—before O.J.—Socrates was tried in 399 BC by a jury that had hundreds of members. Our own jury system evolved from English courts; the Normans who conquered England in 1066 and the Anglo-Saxon kings before them provided early models for the jury system. Juries of noblemen in Normandy decided land disputes—and the duke, or the largest landowner, couldn't act as a judge in his own case. In 1215, the Magna Carta gave English nobles and freemen the right to be tried by a jury of their peers, rather than by the judgment of the king.

> All we see about us, kings, lords, and Commons, the whole machinery of the State, all the apparatus of the system, and its varied workings, end in simply bringing twelve good men into a box.
>
> HENRY PETER, LORD BROUGHAM, "PRESENT STATE OF THE LAW," FEBRUARY 7, 1828

Ninety percent of the world's jury trials take place in America, where the right to trial by a jury of one's peers is fundamental. Juries are "finders of fact"; judges interpret the law and instruct juries accordingly; juries decide innocence or guilt.

There are those who consider twelve-tone music a trial of another sort. Twelve-tone music, or serial music, or tone row, uses all twelve tones of the chromatic scale—the diatonic tones plus the halftones, the sharps and flats—instead of using a few selected notes or chords as the basis or theme for a composition. None of the twelve tones predominates in serial music, and there's no melody as we traditionally understand that concept in Western music. An underlying principle is that no one of the tones can be reused until all twelve have been sounded. The resultant "tone row" of twelve notes isn't necessarily a theme, but is a backbone—an idea that permeates an entire composition. Twelve-tone music is not pitch-centered; there's no home key, so it's not tonal; it is, instead, atonal. Twelve-tone music was

> There are only twelve tones and they need to be treated carefully.
>
> PAUL HINDEMITH

codified by Arnold Schoenberg in the years around World War I, and among his followers were Alban Berg, Anton Webern, and, ultimately, Igor Stravinsky.

There is still a host of twelves—the twelve labors of Hercules, the twelve ribs that keep our lungs and heart protected, the twelve Great Feasts of the Eastern Orthodox Church, and a surprisingly large number of movies with twelve in the title: *Twelve Monkeys, Ocean's Twelve, Twelve Chairs, Twelve Angry Men,* and *The Dirty Dozen.* And Shakespeare's *Twelfth Night.* The United States is divided into twelve Federal Reserve Districts (Atlanta, Boston, Chicago, Cleveland, Dallas, Kansas City, Minneapolis, New York, Philadelphia, Richmond, St. Louis, and San Francisco), and dollar bills have serial numbers beginning with one of twelve different letters—A through L—to represent the Federal Reserve Bank from which the bill originated.

The zodiac is a geometric division of the sky above our heads—that part of it that the sun seems to cross in its daily passage, a belt of sorts that also includes the moon and the planets we can see without telescopes. It's an ancient concept—the Babylonians, as far back as 2000 BC, used the zodiac as a way of visualizing the passage of time. Each of the zodiac's twelve signs represents one-twelfth of the year, linked to a constellation in the sun's path.

The signs and the constellations they're named for don't exactly correspond: the signs are sections of the sky; the constellations are clusters of stars that in three-dimensional space may be thousands of light-years apart. But astrologers believe that our personalities are influenced by the characteristics of the sign we're born under, and the position of the sun, moon, and stars at the moment of birth. The signs are Aries, Taurus, Gemini, Cancer, Leo, Virgo, Libra, Scorpio, Sagittarius, Capricorn, Aquarius, and Pisces. The word *zodiac* comes from the Greek *zodiakos,* "circle of animals," but not all the signs are related to a zoo. Libra, most obviously, is the scales, and is the single inanimate sign in the twelve.

The Asian zodiac is different. The Western concept of time is linear: it progresses in a straight line from the past to the present and stretches forward to the future. In traditional China, time was looked at as repeating over and over according to a pattern. (China adopted the Western solar calendar in 1911, but the old calendar was lunar-based, and is still used for the celebration of the New Year and other festivities.)

One of the ways of recording the cyclical pattern of the years was the twelve animal signs: each year has an animal sign that repeats in twelve-year cycles. According to Chinese legend, the twelve animals had a quarrel because they couldn't decide who should begin the cycle. The gods decided to hold a contest: The animals would have to swim across a river, and the one that reached the opposite bank first would lead off the year; the rest would follow according to the order of their finish. The animals gathered at the river's edge, and at the sign, all jumped in and began swimming as fast as they could. The rat, though, had jumped on the ox's back, and just as the big animal was about to climb ashore on the opposite bank, the rat jumped off, scampered ahead, and won the race. The pig is a slow and lazy animal, and came in last. That's why the rat begins the twelve-year cycle, the ox is second, and the pig is last. The others, in order, are the tiger, rabbit, dragon, snake, horse, sheep, monkey, rooster, and dog.

> **Your letter is come; it came indeed twelve lines ago, but I could not stop to acknowledge it before, & I am glad it did not arrive till I had completed my first sentence, because the sentence had been made since yesterday, & I think forms a very good beginning.**
>
> **JANE AUSTEN,** LETTER TO HER SISTER, CASSANDRA, NOVEMBER 1, 1800

The Japanese tell a different story. Ages ago, on the first day of a new year, Buddha called all the animals of the world to come to him. He said that those who gave him the honor of coming at his call would be given a gift for their loyalty—a year would be named after them. Of all the animals there are in the world, only twelve came, and they arrived in the same order as the animals on the Chinese riverbank. Each animal contributed its traits to its year, and whoever is born in that year has those characteristics, both strengths and weaknesses—the dragon's energy, honesty, bravery, stubbornness, and short temper, for instance. Knowing the year's animal makes possible predictions for the future.

The year 2000 was the year of the dragon, and the years follow in order—snake, horse, sheep, monkey, rooster, dog, boar, rat, ox, tiger, and rabbit, and then dragon again, forever into the future and forever into the past, keeping time and welcoming new babies in that same twelve-year cycle, over and over again.

> **Whoever thinks of going to bed before twelve o'clock is a scoundrel.**
>
> **SAMUEL JOHNSON**

Twelve does that. It seems just that much more advanced than ten. It opens new doors, looks through new windows, makes possible all sorts of ways of thinking. It's the last of the new number names until we reach hundred—leaving a long way to go. It's also the highest number of all the numbers named with just a single syllable, with the magic and splendor—the almost imperial nature—of that one strange, richly consonant sound: twelve.

# 100

## HUNDRED IS A STRANGE WORD

As you might expect given the majesty and magic with which they imbued ten, the Greeks believed hundred to be a marvelous number, the number of perfected perfection, perfect Good. Hundred, after all, is ten times ten, ten squared, ten tens.

> **I think and think for months and years. Ninety-nine times, the conclusion is false. The hundredth time I am right.**
>
> **ALBERT EINSTEIN**

After all these years, we're still impressed by the idea of a hundred. And some of its power is evidenced in the use of the article *a* that so often precedes it when we speak—even when we write. We don't say a ten, a twenty, or a nine—not even a fifty—unless we're talking about, for instance, a one-dollar bill, or one particular card in a deck of fifty-two. A ten of spades. But we *do* say *a* hundred, *a* thousand, *a* million, when we could just as easily say one hundred, one thousand, one million. Hundred is the first of those large and *articled* numbers, the article bestowed, no doubt, because they *are* so large. A hundred is that kind of quantity, not at all run-of-the-mill. A hundred is special. It is, after all, big.

We remain impressed by the idea of a hundred of anything: the first hundred days; a list of one hundred; or even just a century of a hundred years. Hundred, after all, comes perilously close to being our *many* word—it takes a heap of counting to reach one hundred—and some of its power lies in just that idea of *many.* "If I've told you once," mothers say, "I've told you a hundred times." What they really mean is that they've told you too many times to have to tell you again. Will you please *listen?* And when we say, "Hundreds of people are coming to the meeting," what we really mean is a great many.

Still, we think of hundred as a definite, finite quantity, and of course it is—at the same time that it doubles for *many.* A hundred isn't ninety-nine, or a hundred and one, or any other number. It is what it says. But a hundred was once so vast as to be almost unthinkable, just as a trillion and a quadrillion are today. Millionaires are a dime a dozen, and billionaires come along astoundingly often; the national debt numbers in the trillions—and it's all unimaginable. You don't have to define that postbillion number by saying one trillion. Just a trillion will do. In a related way, numbers can be adjectives as well as nouns—think of "the three little pigs," as opposed to "the three," or "the Chicago Seven." In the grammar of our souls, a hundred just *is.*

> **Outside, among your fellows, among strangers, you must preserve appearances, a hundred things you cannot do, but inside, the terrible freedom!**
>
> **RALPH WALDO EMERSON**

Part of that is the unconscious memory of finger counting. Counting to ten on your fingers is natural, even inevitable. Ten times your fingers, though, is a lot of counting, and the numbers that lie even beyond that seem at first to be part of the intangible, indefinable, ineffable *many.* Hundred is a good place to stop, even if it eventually proves only temporary.

Novelist Peter Dickinson writes, in *Tefuga,* about a group of native Nigerians who count by twenties, with their fingers and toes, using assorted names for groups of twenty until they reach one hundred. Then they put a stick on the ground and stand on it to mark their place, and begin counting again with their word for one. Two sticks equals two hundred, and the next number, again, is one.

Maybe that's why adding just one to a hundred seems like an extravagantly enormous quantity. The same thing was true with eleven, and will be again with a thousand, when we get there. Here, a hundred and one of anything is a vast cloud of numbers. That single unit on top of a pile of a hundred seems magical, a number almost beyond our power to count. It's the single straw that brought the poor camel to its knees, the one too many on top of everything else. A hundred and one years is a sentence of doom, an exile for eternity, as if a hundred years might not be enough. A lease of a hundred and one years is virtually forever; a hundred and one dalmations is a whole lot of black-and-white puppies. (Well, of course, it *is*.)

> It signifies not whether a man shall look upon the same things for a hundred years or two hundred, or for an infinity of time. . . . The longest lived and the shortest lived man, when they come to die, lose one and the same thing.
>
> MARCUS AURELIUS, *MEDITATIONS*

## RATIONAL HUNDREDS

The word *hundred* comes by way of a commodius vicus of recirculation from the Latin *centum*. We see *centum* in such words as *century* (one hundred years), or *cent* (one-hundredth of a dollar), or *percentage* (a part of a hundred), but not very clearly in hundred itself. Hundred has a rather violent sound to it, as if it had been brought to us by a horde of invading huns. In fact, it comes most directly from the Aryan word *hundert,* which itself was by way of the Greek *hekaton* and the Indo-European *kmtom*.

The *-red* part of hundred goes back to the Latin *ratio,* "to calculate," or "figure," or "count"; *ratio* also gives us, in English, *ratio* as well as *rational* and, as luck would have it, *irrational.*

The Roman *centurion* was a professional soldier, holding a rank comparable to an American captain, with senior centurions more or less on the level of major, but all sounding a bit like the stereotype of a master sergeant. They originally commanded a *century* of a hundred men, but that number grew larger over time. Centurions trained the men under them, and were reputed to be tough, merciless, and brutal; they themselves fought not only alongside their troops, but in the very front lines, and they suffered heavy casualties for that reason.

Modern centurions are equally tough: there's a British tank, a diesel plane engine, a Cessna airplane, a Buick car, and, on *Star Trek,* a low-ranking commissioned officer in the Romulan Star Empire. In *Battlestar Galactica,* a centurion is a Cylon warrior. Tough guys, all.

The Roman numeral for hundred is C, which was indeed influenced by *centum*—but *centum* isn't wholly responsible. The original sign for hundred was like two inverted parenthesis signs with a vertical line between; gradually, one or the other of the parentheses— ) or ( —was used alone because it was faster to write. The final curve— ( —proved irresistible: it resembled the first letter of *centum,* and C became the sign for hundred.

The Roman numeral for five hundred went through a similar process, evolving from a busy three-line form, with the curve reaching the line before its ends, and with a diagonal slicing through the curve, all tilted slightly to the right. Its next phase was more upright—the tilt disappeared, and the sign sat straight—and then the vertical and the curve got larger while the cross line got smaller, and finally, the cross line disappeared altogether, leaving a well-defined D to stand for five hundred.

Chinese numbers are pictographs, drawn in many different

> **Let a hundred flowers bloom.**
>
> CHAIRMAN MAO, USHERING IN THE CULTURAL REVOLUTION

ways. One, two, and three are usually the standard straight line (horizontal rather than vertical), but the other numbers are more complex. It's a ten-based system, with separate names for the numbers from one to ten, followed by new names for hundred, thousand, and ten thousand. All the other numbers are combined forms of numbers that have gone before, like English with its names for one through twelve, and hundred, thousand, and million.

For us, a hundred is always a hundred, the number after ninety-nine and just before one hundred and one. It was not always so. Before the eighteenth century, there was a small hundred, usually (but not always) a fairly exact one hundred, and a long or great hundred, which was larger—often a hundred and twenty. (Sometimes a hundred was neither: a thirteenth-century note on measurement says, "A hundred of Hard Fish is Eight Score"—a hundred and sixty.) Small or great hundred, by any other name, it's all a way of sliding into large numbers, as if once they get that big, they can't be sure to stay in one place. We can blame the idea of the great hundred on the Teutons who invaded England after the Romans left. Their *hundert* equaled one hundred and twenty. Slippery. The German word for hundred is still *hundert,* but now it is precisely, exactly, and always one hundred, right on the nose.

## HUNDREDS AT WAR

Even in mythology, hundred is a large number. The Hundred-Handed Ones (in Greek, the Hecatoncheires) were huge creatures who were the offspring of Gaea, who was the earth, and Uranus, the god of the sky and of heaven. Their other children were the Titans, of whom Cronus was the leader, and the one-eyed Cyclopes.

In one version of the Hecatoncheires myth, after they were born, Uranus saw how monstrous they were and pushed them back into Gaea's womb, thus imprisoning them in the earth. She was not happy about this—it hurt!—and asked the Titans for help and vengeance. Cronus castrated Uranus, and where his blood fell onto the earth, the three Furies arose—they were the goddesses of revenge. The severed testicles of Uranus fell into the sea, where they produced a quantity of white foam, out of which Aphrodite, the goddess of love, was born, rising up on Botticelli's half shell—a bit too late though, for heaven and earth, Uranus and Gaea, had already been separated and have never again been reunited. The Hecatoncheires, meanwhile, are said to represent the great forces of nature, like earthquakes and tidal waves. Not a happy story, but few of them are.

There are Greek myths, too, about Argus, who had a hundred eyes and could see nearly everything around him. It happened this way: Zeus, the ruler of the Olympians, was the husband of Hera, their queen. Hera's first priestess was named Io. Unfortunately, Zeus fell in love with Io and had a passionate affair with her. Even more unfortunately, Hera found out. Disaster threatened. In order to keep Io safe, Zeus changed her into a white heifer. But Hera was nobody's fool; once again, she found out, and commissioned the hundred-eyed Argus to guard the little white cow who had been Io, and to let her know if there was any sign of Zeus in the vicinity. In the next installment, Zeus had the god Hermes kill Argus. In tribute to her guard, Hera transplanted his hundred eyes onto the tail of her peacock, which is why peacocks have all those eyes—all those colors—in the feathers of their tails.

> An hundred years
>   should go to praise
> Thine eyes, and on
>   thy forehead gaze;
> Two hundred to
>   adore each breast;
> But thirty-thousand
>   to the rest.
>
> ANDREW MARVELL, "TO HIS COY MISTRESS"

That was the end of Argus, but Hera wasn't done. She sent a gadfly to irritate Io. Trying to get away from the fly's torment, Io crossed what we now call the Ionian Sea, yes, named after her, and then swam the Bosporus Strait (literally, the "Ox-ford," where the ox, who was Io, crosses). At last, she reached Egypt, where she was able to rest and was restored to her original form. She eventually became the mother of Epaphus, who was to be a king of Egypt.

There are historical—as opposed to mythical—hundreds, most notably the Hundred Years' War, and Napoleon's hundred days. The war began in 1337 and ended in 1453, which is really a hundred and sixteen years, but, given the power of hundred, we've come to think of it as a nice round number. The war (actually wars, because there were intervals of peace) was a series of armed conflicts between Europe's two greatest powers at the time, England and France, and had to do with England's claim to French territory and the French throne.

The roots of the war go back as far as 911. Charles the Simple (son of Louis the Stammerer, and eventual successor of Charles the Fat) was king of France; he allowed Vikings to settle in the part of his kingdom we now call Normandy (after the Norsemen). By the time they invaded England, the Vikings were known as Normans. They were led by William the Conqueror (the Duke of Normandy)—this was the Norman Conquest of 1066. They defeated the Anglo-Saxons, and William became the king of England, and the Normans ruled both Normandy and England for the next 150 years. In 1216, the English Normans lost most of their continental possessions to France. By the fourteenth century, most English nobles were descendants of the Anglo-Normans; they spoke French, and remembered when they ruled Normandy as well as

**As she frequently remarked when she made any such mistake, it would be all the same a hundred years hence.**

**CHARLES DICKENS,**
*MARTIN CHUZZLEWIT*

England. Their goal was to reconquer the land in France they felt belonged to them—a rich territory that would give England great power and not incidentally increase the wealth and prestige of the English nobility.

In a large sense, the war changed medieval society; it certainly changed the way wars were fought. France was the larger country, both in land and in population, and it still embodied the chivalric tradition. Wars were fought by mounted knights, noblemen whose goal was to dismount enemy soldiers and hold them for ransom. The tactics of the English, smaller in number, were to wound and kill. Their weapons included the longbow, and they were the first to use gunpowder (invented in China) in battle on European ground; their army included archers and ground soldiers. The nobility was no longer the deciding factor on the battlefield; peasants had access to weapons, and thus to power and prestige. Soldiering wasn't limited to feudal lords; now, in addition to peasants in the ranks, there were paid mercenaries—and after the war, standing armies for the first time since the Romans. The age of chivalry was coming to an end.

> A thought is often original though you have uttered it a hundred times.
>
> OLIVER WENDELL HOLMES, *THE AUTOCRAT OF THE BREAKFAST TABLE*

The war's final turning point came in 1429 with the Joan of Arc–inspired French victory at the siege of Orléans. Thousands of lives had been lost, but in the end France emerged as a united nation under an absolute monarchy, rather than a collection of fiefdoms, and England relinquished all claims to French territory and turned to the sea, eventually becoming a naval power instead.

Napoleon's hundred days are another chapter in French history. They began with his arrival at the Tuileries, in Paris, on March 20, 1815, after his escape from Elba, where he had been exiled. He resumed his title as emperor, and his army rallied behind him. It was a brief reign; on June 28 of that year Louis XVIII was restored to the throne of France, and Napoleon was banished again, and this time finally—to Saint Helena, where he died of stomach cancer a few years later. The hundred days include the Battle of Waterloo, which has become a synonym for crushing defeat. Napoleon was routed at Waterloo by the combined troops of several European countries. It's estimated that on that one day, June 18, 1815, forty-seven thousand men died or were wounded in an area of a few square miles; many more had been killed or injured in the three days before that epic battle. When it was all over, the balance of power in Europe had shifted, boundaries were redrawn, and French domination of Europe, under Napoleon, had ended.

> **How can anyone govern a nation that has two hundred and forty-six different kinds of cheese?**
>
> **CHARLES DE GAULLE,** QUOTED IN *NEWSWEEK*, OCTOBER 1, 1962

## HUNDREDS AND HUNDREDS

We speak of the first hundred days of American presidencies, and that tradition goes back not to Napoleon, but to Franklin Delano Roosevelt. Much of the New Deal legislation, the hallmark of his presidency and the saving of America from the worst effects of the Great Depression, was enacted between March 9 and June 16, 1933—ninety-nine days that we again round off for the power of one hundred.

There's more before hundred is over: antiques, legally and traditionally, should be at least a hundred years old to be genuine. Hundred-percenters were originally thoroughly patriotic (not to say jingoistic) Americans in the 1920s and 1930s. In India and Israel, 100 is the police telephone number, and in the UK, it's the number for the operator.

Hundred was once the name of an administrative division of

land that divided large regions into small units capable of sustaining a hundred families each. The idea goes back to a Germanic system described by Tacitus in AD 98 and brought to England by the Saxons in 613. It may have been linked to the amount of land administered by hundred-man subdivisions of the Teutonic army. The division of shires into hundreds lasted until the nineteenth century. (*Sheriff* is a shire-descended word.) Sweden, Norway, Denmark, and Finland once divided their countries into similar hundreds. The concept traveled to America; counties in Delaware, New Jersey, and Pennsylvania had areas of hundreds in the seventeenth century. In Delaware today, hundreds are still used, but only for real estate title descriptions.

In China, hundred has an extra reverberation. There, days were once divided into a hundred sections, each called a *k'o*. The sections ran from midnight to midnight, and were kept track of with water clocks and incense sticks that burned at a measured rate. During droughts, water clocks couldn't be used. An imperial official invented "the Hundred K'o Incense Seal"—a series of grooves arranged like a maze. Incense was spooned into the grooves and lit at one end. It burned at a slow and steady rate: it took each groove one-hundredth of a day, a *k'o,* to burn. You could tell the time by seeing how much incense had burned.

This is a concept of time as well as of number. There are so many places on our earth where time and numbers cross: on the pages of our calendars, in the turnings of the moon, and in the naming of our days as our planet travels around the sun. We have only to keep track, whether through the flooding of the Nile, the burning of the *k'o,* or the ticking of the clock, to know that time is fleeting and our days are numbered. Each one passes slowly; but together, by the hundreds, they slip between our ten fingers.

**There is only one religion, though there are a hundred versions of it.**

GEORGE BERNARD SHAW, *ARMS AND THE MAN*

# 1,000

## THOUSAND IS A SWOLLEN HUNDRED

When we come to the word *thousand,* we also come to the first time the letter *a* has been used in spelling out a number. Not in one, two, three, four, five, six, seven, eight, nine, ten, eleven, twelve, or hundred is there a single *a;* not until we reach thousand.

There are all sorts of similar facts for the trivia-minded among us. The first number name to use all five vowels in any order is one thousand five, and the first to use the vowels in their a-e-i-o-u order is one thousand eighty-four. There are fourteen *e*'s in the number names from one to twelve—and eleven has three of them. The first name with a *b* in it is—are you ready?—one billion. That's a long way to go. And before I stop myself, there are no number names at all with the letters *z* and *k* in them. (*Q* is for quadrillion.)

A thousand, complete with both its letter *a* and its article *a* as well, is a great and glorious number, and for a long time it was as high as a number needed to be. But when you get right down to it (or up to it), it's just an extension of one hundred, and one hundred is just an extension of ten, and ten, in its turn—now somewhat humbled—has grown out of one. It all comes back, always, to one. The difference is that once you've

> A little one shall become a thousand, and a small one a strong nation.
>
> ISAIAH 60:22

gone to one hundred, a thousand isn't such an incomprehensible leap. It's merely more of the same.

Thousand also marks the moment when we start separating numerals with a punctuation mark. In the United States and Great Britain we use a comma to differentiate between hundreds, thousands, and millions. In many other countries, a dot is used. The American 1,234,567,890 is much easier to read or even just to think about than 1234567890. In other places, that would be 1.234.567.890 or sometimes even just 1 234 567 890. (Swiss German uses neither a dot nor a comma, but instead, an apostrophe: 1'234'567'890.) When we're dealing with a decimal point, the difference is backward: in the United States, decimal point numbers, no matter how large, are written with no punctuation. In other places, every trio of post–decimal point numbers is separated by a comma (14.320,1 as opposed to 14.3201). But—there's always a but—some countries indicate the decimal point not with a dot but with a comma. Are we back where we started? The way the other guy does it looks strange to the rest of us. But no matter how it's done, it goes back to the old rules: given our limitations, it's very difficult for human beings to count at a glance groups that are larger than three or four.

Whether they are separated by a comma or just lumped together, the mind reels at numbers with seemingly endless rows of zeros, but even before the idea of zero existed, there were limits to how large a number could be imagined—even to how large a number was needed.

Those cultures that did need large numbers often used the idea of multiples of ten to write them. The Egyptians had hieroglyphic signs only for one, ten, and multiples of ten. One, their number of unity, was a single vertical line. Ten looked like an upside-down U; hundred was a small, left-facing spiral. Thousand was like a lotus flower. Hundred thousand was a tadpole. Ten thousand was much simpler; it looked like a slightly bent finger. Million was a kneeling man with his arms stretched up to the sky. To write all the numbers between, the signs were repeated until they added up to the desired number. Four hundred meant four spirals; seven thousand meant writing seven lotus flowers.

> The creation of a thousand forests is in one acorn.
>
> RALPH WALDO EMERSON

Did the signs have meanings beyond the numerical? The vertical line for one has been used thousands of times over the aeons; it's a natural symbol—one raised finger. The Egyptian upside-down U for ten could be two joined hands (like the Roman X, only there one hand is upside down, or the arms are crossed). The spiral and the lotus, according to Georges Ifrah, are probably representations of words that sounded like those images, though their meanings may have been different. (In classical Chinese writing, he notes, the pictograph character for man is used to indicate a thousand, because the words *thousand* and *man* had the same sound.) The Egyptian bent finger for ten thousand is probably a remnant of finger counting. The tadpole, for hundred thousand, might be remembering the vast number of newborn frogs in the Nile in springtime. At a million, the kneeling man might be in awe of such a huge number (perhaps he's praying that there will be nothing higher)—or he could be a priest looking at the stars in the night sky. (Ifrah suggests that he represents a genie holding up the vault of heaven.) The same sign means not only an abstract million, but also a more specific, if still unthinkable, million years—or eternity.

The Greeks stopped writing new numbers when they reached ten thousand, and gave that enormously large number the name *myrioi* (from which *myriad*)—once again, the eternally elusive *many*. The written sign for ten thousand was an M, from *mu,* the first Greek letter of *myrioi.*

The Romans went higher than the Greek ten thousand, but only as far as hundred thousand. They could imagine numbers

> I have never quite grasped the worry about the power of the press. After all, it speaks with a thousand voices, in constant dissonance.
>
> ERIC SEVAREID

higher than that, but expressed them as compounds of numbers that had gone before. One of their early signs for thousand was a circle with a vertical line through it, as if the circle were a large multiple of one. Later, the single line was replaced by an X for ten. Later still, the inner lines became a vertical cross, like a t, but still circled. There were many variations, but eventually the letter M—with which the Latin *mille,* thousand, begins—replaced all the circles, curves, and lines that had gone before.

It's to the Latin *mille* that we owe such words as *millennium*—a thousand years—and less grandly, *mile,* because a Roman mile was a thousand paces long. A millimeter was a thousandth of a meter, as a *milli*-anything is a thousandth of an anything. A kilometer—a thousand meters—is from the Greek, *khilioi,* thousand. The Greek number sign for thousand was X, from the first letter *chi,* in Greek, of *khilioi.*

The English word *thousand* itself has nothing to do with either *mille* or *khilioi*. It's related instead to the Latin word that indicates largeness. *Tumere* means to swell, or to be swollen. Also from *tumere: thigh* (the swollen part of the leg), *thumb* (the swollen finger), and most obviously, *tumescent* (becoming swollen), *tumid* (inflated), and *tumor* (a swollen mass of new tissue). A thousand, then, is a swollen hundred.

The same sense is in the Sanskrit *tawas,* "strong," and the Aryan *tus*—a "strong hundred." For a long time, thousand meant a multitude (think *many*), and for a long time, one thousand was not the number after nine hundred and ninety-nine; it was an indefinite sum, too large to count, way out there with *many.*

## MANY THOUSANDS

The tradition of thinking of a thousand as *many* is long and honorable. The night has a thousand eyes. A picture is worth a thousand words. (Well, maybe.) Mahler's Eighth Symphony is called the Symphony of a Thousand because of the enormous number of instrumentalists and singers it calls for. Lon Chaney Sr. was the Man of a Thousand Faces because of the many disguises he used.

"A thousand years are before you like a day," the psalmist sang. In English, we wish our friends and relations many happy returns on their birthdays, but the Chinese wish theirs a thousand springs. Same thing.

The German word for daisy is *Tausendschönchen*—"a thousand little beauties"—and their word for the English word *centipede* (from the Latin for "hundred feet") is *Tausendfüssler,* "creature of a thousand little feet." (Germans can definitely count; their bug isn't bigger, it's just a swollen hundred.) (There is indeed a millipede, from the Latin for thousand feet, and in the tropics it not only has many feet, it can also grow to a length of twelve inches. Ick.)

The Thousand Islands are a group of islands in the Saint Lawrence River between Canada and the United States. Some of them are little more than sandbanks, and their total number, though it's well over a thousand, is inexact—more like *many.* I've seen it numbered at 1,500, 1,700, and 2,100. The 1000 Islands Information Center Web site uses the numeral rather than the word, and in its very first sentence uses that number eight times.

Thousand Island dressing, according to the Herald Hotel in Clayton, New York, did indeed come from the Thousand Islands—it was originally created by a local fishing guide's wife. His clients loved her salad dressing, and one of them gave the recipe to the owner of Manhattan's Waldorf-Astoria. He told the hotel's famous chef to put the dressing on the menu, and Oscar of the Waldorf

> The thought of two thousand people crunching celery at the same time horrified me.
>
> **GEORGE BERNARD SHAW,** EXPLAINING WHY HE HAD TURNED DOWN AN INVITATION TO A VEGETARIAN CELEBRATORY DINNER.

thus was credited with devising it. There are hundreds—if not thousands—of recipes for Thousand Island dressing, and many other claims for its origin. Mostly, it's Russian dressing, made with mayonnaise and ketchup (or, sometimes, chili sauce) and a variety of ingredients that usually include sweet pickles, hard-boiled eggs, and chopped pimiento.

## AND ONE

Scheherazade and her *Thousand and One Nights* are another example of the power of one when it's added to an already large number.

*The Arabian Nights*—the other title for Scheherazade's stories—is a collection of fairy tales, romances, legends, fables, parables, and adventures; the most familiar are the ones about Aladdin, Ali Baba, and Sinbad the Sailor. Their framework is the story of Scheherazade herself. Her father was vizier to the Persian king Shahryar, who killed his wife because she had been unfaithful. Then, in his fury, he decided *all* women were probably unfaithful, so he ordered his vizier to provide him with a new wife every morning, and he had his new wife killed each night. The vizier's daughter, Scheherazade, begged her father to let her marry the king before there were no women left at all—she had a plan, she said, and he reluctantly agreed. On the night of their wedding, she told the king a story—but she refused to finish it until

> Each Morn a thousand Roses brings, you say;
> Yes, but where leaves the Rose of Yesterday?
>
> **EDWARD FITZGERALD,** *THE RUBÁIYÁT OF OMAR KHAYYÁM*

the next night. The king wanted to know how the story ended, so he let her live one more day. But the next night, when she finished her first story she immediately began another—and did the same thing, refusing to end the new tale until the following night. This went on for a long time—not for just a thousand nights, but for a thousand *and one* nights—until finally the king abandoned his awful plan, and they lived happily and, one presumes, faithfully ever after.

A thousand and one nights sounds like forever, much longer than just a thousand nights would be, but for the record, it's about two years and nine months. The first written record of *The Arabian Nights* appeared in the tenth century, and there's disagreement about whether or not there were exactly 1,001 stories then. Some say that there were only 480 stories (but who's counting?) and that the remaining 521 have been added in the years since.

There are scholars who say that the core of the stories was a Persian collection of folktales called *Thousand Myths,* followed by an Arabic compilation, *A Thousand Nights,* in about AD 850. The extra night—to make a thousand and one—appeared for the first time in the Middle Ages; the framing story of Scheherazade may have been added in the fourteenth century, although it probably existed earlier. The stories don't have a single author—like Western fairy tales, they were told and retold orally, shaped and reshaped through the generations, from parent to child, from friend to friend. In any case, for a long time, a thousand represented the idea of infinity—a thousand and one was almost beyond imagining. A legend sprang up that whoever read the whole collection of stories would go mad.

## AND MANY

In terms of counting, for a long time a thousand was sufficient. More swollen than that it wasn't necessary to be. But, as

Scheherazade knew, there is always one more, and one more, and one more. Ten thousand was once a quantity almost beyond thought. The Greek *myrioi* meant countless before it meant ten thousand. Ten thousand was such a vast number that it was almost impossible to conceive of needing to count that high. In the same way, today, we think of the stars we see in the sky over our heads at night as being countless, so many that they're virtually impossible to count. But they *have* been counted, one by one, and there are only a few thousand visible to the naked eye—estimates vary from one thousand to about six thousand. At most, then, six thousand stars, crowded in the dark sky on a moonless night: we look at them, and we think they're beautiful, and unknowable, and countless. Did someone say *many?* Perhaps that's why in China, ten thousand years once meant immortality—may you live ten thousand years, may you live forever—and ten thousand things meant everything that exists.

> A telescope will magnify a star a thousand times, but a good press agent can do even better.
>
> FRED ALLEN

The Persian army once included an elite corps known as the Ten Thousand Immortals. They were the king's bodyguard troops, and if anything happened to any of them, they were immediately replaced—a kind of serial, even if not personal, immortality, and a constant ten thousand. Among their special privileges as an elite guard: They were allowed to take concubines with them on marches. And their dress and armor glittered with gold.

As difficult as it once was to imagine ten thousand of anything, Archimedes, the great Greek mathematician, found a way to work with higher numbers. Archimedes was not only a mathematician—and considered to be one of the greatest of all time—he was also a physicist, inventor, astronomer, and philosopher. He was born in the Greek colony of Syracuse, on the island we now call Sicily, and is said to have invented a variety of amazing war machines to keep Roman invaders away during the Punic Wars. It's Archimedes who said, "Give me a place to stand and I will move the earth," about levers and pulleys. He was said to have repelled one Roman attack by focusing mirrors (or highly polished shields) to reflect sunlight on the ships so they would catch fire and burn, though there's disagreement about whether or not that's really possible. At the very least, the reflected light was blinding. The story about his death is that he was working on a problem during the long siege; when the Roman forces won the city, Archimedes didn't realize what had happened. According to one story, Archimedes was drawing diagrams in the sand when Roman soldiers arrived. "Don't step on my circles," he said, and a furious soldier drew his sword and killed him on the spot. Plutarch told three different stories about the death of Archimedes—but the line about his circles seems to have lasted the longest.

> Thousands of candles can be lighted from a single candle, and the life of the candle will not be shortened. Happiness never decreases by being shared.
>
> **BUDDHA**

It was Archimedes who, as a mathematician, decided to count every grain of sand in the universe, beginning with the grains of sand on the beaches of Sicily. He added to that total all the sands of all the lands of the world, known and unknown, and then decided to determine how many grains of sand there would be in the whole world, and then if the world were filled with sand—okay, how many grains of sand there would be if the entire universe held nothing but sand—the number of grains that would be needed to fill a sphere whose diameter would equal the distance from earth to the stars.

> One is too many and a thousand is not enough.
>
> **PROVERB**

This is how he did it: He figured out how many grains of sand there would be in a lump the size of a poppy seed, and how many poppy seeds there would be in a row as wide as a finger. He knew there were ten thousand finger widths in a Greek unit of length roughly equal to a tenth of a mile and then figured out how many finger widths would fit in the universe—measured then as the sphere of the fixed stars. All of this he did without using zeros—without the *idea* of zero—as well as without having a number larger than ten thousand.

He used, instead, the concept of myriad, of ten thousand. He began with a myriad myriads and called that a number of the first order. Then he went on to numbers of the second order, and the third order—all the way to numbers of the eighth order. Then he put the orders together to make numbers of the first period, and ended with "a myriad-myriad units of the myriad-myriadth order of the myriad-myriadth period." In our words, that would be ten followed by a staggering number of zeros. Some say that the last number of his first period is 1 followed by 800 million zeros, and the final number has ten to the eighth power times as many. If this is true, it would mean that if you had begun reciting the digits of his final number at the moment of the big bang, when the not-so-sandy universe was created, and if you had recited one digit per second from then until now, you would not yet be finished.

> **Tao gave birth to one,**
> **One gave birth to two,**
> **Two gave birth to three,**
> **Three gave birth to all the myriad things.**
>
> ***TAO TE CHING***

It was Archimedes, who lived from about 287 to 212 BC, who discovered the principle that a body immersed in fluid is buoyed up by a force equal to the weight of the fluid displaced by the body. According to the misty voices of history and legend, he was in his bath when he made his discovery, and he is supposed to have risen from the tub and run naked through the streets, shouting "Eureka!" ("I have found it!") What could he have said when he arrived at the final number of grains of sand in the universe? A myriad myriads of eurekas? Not sufficient, surely.

# 1,000,000

## MILLION IS NOT YET FOREVER

To think, it all started with one! One plus one plus one, a million times. If we went backward, if we subtracted, a million minus one minus one minus one, it would all end with one as well—unless we also eliminated that final one, that last solitary number, brave and solid, strong and sure, so that we ended with nothing, with zero: the Garden before Adam, the universe empty, the cipher before the circle enclosed it.

> **Be embraced, O millions!**
> **This kiss for the whole world!**
>
> **FRIEDRICH SCHILLER,**
> **"ODE TO JOY"**

A million of anything is so many that we can barely begin to comprehend it. And yet million is just the beginning of the unimaginably large numbers—the myriad myriads of the first order of the first period.

It's also another in the endless variations of *many*. I wouldn't do that in a million years, we say. That fellow is one in a million. I'd walk a million miles for one of your smiles, according to the song. And while being a millionaire isn't the same big deal it used to be, it still means someone has a lot of money. A dollar may not buy what it used to, but a million dollars is more than enough to pay the rent. In 2005 in the United States, there were over eight million households with a net worth of over a million dollars—not counting the worth of the primary home.

The concept of a million (of anything) is not new. The ancient Chinese had names for big numbers as early as 211 BC, before the Qin dynasty. *Yi* was both ten thousand and one hundred million, and *zhao* was million and used again for thousand million. Eventually, changes were made, eliminating millions of confusing possibilities.

The Egyptians, remember, had the kneeling man with his arms raised, either in supplication (Please! No more!) or in awe; or perhaps he was just holding up the sky with its millions of stars.

The Romans had ways of writing numerals for very large numbers, but they had no names for any number above a hundred thousand. Pliny the Elder wrote that the Romans of his time used the words *decies centena milia,* ten hundred thousand, to mean a million. The Greeks had stopped using number names at ten thousand, *myrioi*.

*Million* begins with the Latin word for thousand, *mille*. The new word was coined in the early fourteenth century by Italian bankers, who found they needed a name for the large numbers they were dealing with. They used *mille* (Italian and Latin both) and added the Italian suffix *-one,* meaning big or great, onto its end. Adding *-one* to *padre,* father, for instance, gives *padrone*—a great father, and thus a patron, master, or landlord. Adding it to *sala,* room, gives *salone*—sitting room, lounge, or even saloon. *-One* makes million a big thousand—just as thousand was a swollen hundred.

At first only bankers used the new word; to everybody else, it was just a very large number, the kind you don't need to have handy around the house because nothing around the house needs to be counted that high. But the banking world did need the number; it stuck, and it traveled.

> But words are things,
> and a small drop
> of ink,
> Falling, like dew, upon
> a thought, produces
> That which makes
> thousands, perhaps
> millions, think.
>
> GEORGE GORDON, LORD BYRON,
> *DON JUAN*, CANTO III

And it grew. By the end of the fifteenth century, a million was no longer enough. *Bi,* from the Latin for two, was used to make billion. And thus trillion, and all the rest. There's no zillion or jillion or gazillion, though there certainly could be; all of the -illions are logical; it's just that they seem so unlikely.

Strangely, and confusingly, there isn't universal agreement on exactly how many a billion is. In the United States and many other English-speaking countries, a billion is a thousand millions. In Great Britain, one billion used to be a million millions, and in many other countries it still is. (The word for the American billion (1,000,000,000) in Great Britain was a *milliard.* Confusing? Yes.) In 1924, H. W. Fowler's important book, *A Dictionary of Modern English Usage,* called for the British to adopt the American system—but it took a half century before that began to happen. In 1974, the British government announced that from then on, in its reports and statistics, one billion would mean 1,000,000,000, just as it does in the United States. The *Times* of London style guide now also speaks of nine zeros plus one in the same way.

In the rest of the world, some countries swing one way in the naming of billions, and some swing the other. Brazil and Greece, for instance, count the way Americans always have; many European countries, from Norway to Spain, count the way the British did before 1974.

For Americans and many other English speakers, the prefix before the root *-illion* tells how many sets of three numeral places are in the written numeral, *not counting the first three,* which go up to 999 and are included in *mi-*. For one billion, there are two sets of zeros that count the thousands and the millions—plus the standing three that count the hundreds: 1,000,000-plus-000.

For others, the Latin prefix tells what power of a million the number is. A billion in those countries is a million to the second power ($1{,}000{,}000^2$), so a billion is equal to a million times a million, or 1,000,000,000,000. BIG difference. And it gets bigger as the zeros go along. Mathematicians and scientists, as luck would have it, have no choice but to speak of numbers this large; they tend to use mathematical notation (indicating powers of the number) instead of word names so that everybody, no matter where they live or what language they speak, will know the number that is meant. In those terms, in the United States, a billion is ten to the ninth power; in Great Britain pre-1974, it was ten to the twelfth power.

How large, then, is 1,000,000,000? A billion seconds ago the parents of a middle-school child were themselves in elementary school. A billion minutes ago the Roman Empire had not yet fallen and Jesus was still alive. A billion hours ago we were in the Stone Age. A billion days ago an apelike creature considered the African savannas his home, and had not the slightest notion of either evolution or intelligent design. A billion months ago dinosaurs were thriving. A billion years ago multicellular organisms appeared on earth for the first time.

To say all that differently, a billion seconds is almost 32 years. A billion minutes is about 1,900 years. A billion hours is 114,000 years. A billion days is 2.7 million years. A billion months is 83 million years. And a billion years—well, the universe is thought to be about 13.7 billion years old.

## MONEY, MONEY, MONEY

There may be many more billionaires than there used to be, but a billion dollars is a lot of money. *The World's Billionaires* is the annual list of the world's wealthiest people in *Forbes* magazine. It's not a comprehensive list—it doesn't include private fortunes that are sheltered from public examination—so there are really many more billionaires than are listed. Even at that, the list has grown. When *Forbes* did its first list, in 1986, there were 140 billionaires on it. Twenty years later, there were 793, more than five times as

> People compose for many reasons: to become immortal; because the pianoforte happens to be open; because they want to become a millionaire; because of the praise of friends; because they have looked into a pair of beautiful eyes; for no reason whatsoever.
>
> ROBERT SCHUMANN

many. Their average net worth, in 2006, was $3.3 billion; their total worth, $2.6 trillion, up 18 percent since just the year before.

The youngest billionaire is Germany's Albert von Thurn und Taxis, who is twenty-three. The oldest on the 2007 list is America's John Simplot at ninety-eight. He made his money from potatoes and microchips. Separately, one assumes. There are several on the list whose age isn't known.

In 2006, 371 billionaires lived in the United States. Next on the list was Germany, with 55. Russia had 33, Japan 27, the United Kingdom 24, India 23, Canada 22, Turkey 21, Hong Kong 17, Brazil 16, France and Italy 14 each, Saudi Arabia 11, Mexico and Spain 10 each, China, Malaysia, Sweden, Switzerland, and Israel, 8 each, Australia 7, Taiwan and the United Arab Emirates, 5 each, and Singapore, South Korea, Ireland, the Netherlands, and Norway, 4 each. Eight countries had 3 each, nine had 2 each, and six 1 each. Thirty-nine billionaires lived in New York City; 20 lived in Moscow, 19 in London, 12 in Paris, and 11 in Tokyo. The list didn't include any city with fewer than 2 billionaires. There were too many of them.

The single person with the most billions? Bill Gates, with fifty-six. (2007 marks his thirteenth straight year at the top of the list.) He's on another list, too, of the richest Americans over the years, where he joins two Astors, two Vanderbilts, Andrew Carnegie, John Rockefeller, Jean Paul Getty I, Howard Hughes, and Warren Buffett.

> I am opposed to millionaires, but it would be dangerous to offer me the position.
>
> MARK TWAIN

In Britain, the London *Times* has published an annual *Sunday Times Rich List.* It isn't limited to British citizens—anyone who works or lives mostly in Britain is eligible, thus excluding Rupert Murdoch, owner of the *Sunday Times,* among many other newspapers. There are 1,001 men on the list and 81 women; of those, 250 inherited their money; the rest came by it through other means. Sources of wealth include land, property, finance, industry, retailing, food production, construction, computers, mobile phones, media, hotels, music, cars, pharmaceuticals, transportation, and—ta da—weight-loss products. A hundred and twenty-five of those on the list are titled, and the Queen is first.

The *Forbes* list for 2006 included only 78 female billionaires. The magazine noted that although there are ten more women on the list than there were the year before, "only 6 are self-made."

**A billion here and a billion there and pretty soon you're talking big money.**

EVERETT M. DIRKSEN,
U.S. SENATOR FROM ILLINOIS, 1950–1969

## -ILLIONS AND -ILLIONS

After the first millions and billions and trillions have gone by, there are still more. We have quadrillion, quintillion, sextillion, septillion, octillion, nonillion, and decillion. The -illion numbers go all the way through the teens (undecillion, duodecillion, tredecillion, quattuordecillion . . . ) to vigintillion, which is 1 with 63 zeros after it. 1,000,000,000,000,000,000,000,000,000,000, 000,000,000,000,000,000,000,000,000,000,000.

There are names for larger numbers, but spelled-out references to them are rare; they aren't found in most dictionaries. The standard way to refer to most of these numbers is to use the mathematical notation showing powers of ten, and when words are needed the powers of ten are spelled out—as ten to the sixty-third power—which, again, eliminates the confusion between ways of counting a billion and all the numbers beyond. Vigintillion, when it's needed, is shown as $10^{63}$, which is a lot of zeros any way you look at it. To go a bit further, centillion is a nice curiosity piece—it has 303 zeroes ($10^{303}$)—but we aren't there yet. Soon.

After counting all those zeros, you might think this is a ridiculous question, but let's ask it anyway: What happens if we keep adding one more?

# And One More

## THE MANIFEST DESTINY OF NUMBERS

There's always one more. It's as simple as that, as obvious as that. It's human nature to count—beginning with ourselves as one, and the nearest as two, and going on from there—and it has become human nature to keep counting, to number not just what we can see, but also to go past that to what we can predict, and ultimately, to count what we can imagine.

> To see a world in a
> grain of sand
> And a heaven in a
> wild flower,
> Hold infinity in the
> palm of your hand,
> And eternity in an
> hour.
>
> **WILLIAM BLAKE,**
> "AUGURIES OF INNOCENCE"

We like to think of numbers as *there,* as always having been there, having arrived complete, from one to forever, all at once, not with a thunderclap but perhaps with a small announcement: One, two, three, go. And we did go, or so we think. We learned to count just as soon as we saw that the numbers were here.

Except that's not how it happened.

We learned our numbers slowly—one by one, in different times at different places, and in different ways. We always started with one—how else?—but when we needed more than one (as we almost immediately must have) we reached out, in all those places and times, to two. Beyond two was a series of leaps, to three, to four, and to a magical *many*—and we did that, leaping and stretching and needing to know more, too many times to number, but in the numbering, we learned our numbers.

Counting grew this way, slowly, from piles of pebbles that equaled one by one what we needed to remember, however briefly. After pebbles we learned to make notches on a stick or a bone, and then we went to a sand table, an abacus, and, in our time, to a calculator and now a computer and networks of computers.

Along the way, we slowly realized that there is always one more, so we slowly found another number, and another way to name it.

That's what googol is. It's a vast number, one with a hundred zeros after it, with a made-up name, one more than the last number, one less than the next. It's the new number of our own time (as nine was the new number in another era). Googol was bound to happen, in this age of technology and incredibly fast computing, in this time of the exploration of vast distances and immense space, outward and inward, when the edges of the universe recede and stretch and curl away, and we follow after, counting and measuring, and marking the distance from what we *see* is there to what we believe *must* be there. We explore neutrons and particles, atoms and molecules, charm and quark, in one direction, and in the other we reach to the last galaxy we can see—and beyond. We redefine the planets. Googol is a number for our age.

> When Moses was alive, the pyramids were a thousand years old... Here people learned to measure time by a calendar, to plot the stars by astronomy... Here they developed that most awesome of all ideas—the idea of eternity.
>
> WALTER CRONKITE, *ETERNAL EGYPT*, ON CBS TV, JUNE 28, 1980

For all of that, although googol, with its hundred zeros, *is* a number (and an even number at that), it's really as much an exercise as it is a numeral, a way of imagining something vast, almost unthinkable. We tamed the wildness of the first forests with the name of one, and now we tame the universe with the names of numbers so large, with so many zeros, that if we wrote them down, if there were sufficient paper and ink, and time to write them, they wouldn't fit within the universe we're numbering.

The BBC's h2g2 page on the Web puts it this way: Imagine a tiny silver ball, about three millimeters in diameter, made of sugar, used to decorate a wedding cake. If we hollowed out the earth and filled our entire planet with tiny silver balls like those on that cake, the number of icing balls we'd have is six times ten to the twenty-eighth power—6 with 28 zeros after it. Way too small. A sphere the size of the solar system filled with sugar balls? Five times ten to the forty-sixth power. Not big enough. A sphere the size of the galaxy? Two times ten to the seventieth power. Not there yet. The universe? The whole universe? Filled with tiny silver balls? Two times ten to the eighty-sixth power. Sticky, but nowhere near enough. Here's how: Fill the universe with tiny silver sugar balls, and then one second later, empty it out and replace all the balls with another identical set. Now do this again—and do it every second for four thousand years. "The total number of silver balls is a googol. And that's a load of balls!"

## NAMING A NUMBER

The name of this new number of our time wasn't based on the name of a previous number with an added suffix, like -teen, or -ty, or -illion. It's new in the same hasn't-been-used-before way that all the numbers up to twelve are, and after twelve, hundred, thousand, and million.

*Googol* is a totally made-up word. It isn't based on anything, not Sanskrit, Greek, or Latin, it doesn't derive from anything, and it has no relationship to anything, except perhaps the universal language of silly sounds.

> [A googol] is more than the number of raindrops falling on the city in a century, or the number of grains of sand on the Coney Island beach.
>
> **EDWARD KASNER,** QUOTED IN HIS OBITUARY IN THE *NEW YORK TIMES* IN 1955

The word *googol* was used for the first time in *Mathematics and the Imagination,* a popular math book by Edward Kasner and James Newman, first published in 1940 and still available. The book tells the story of the invention of googol by Dr. Kasner's nine-year-old nephew, Milton Sirotta. Dr. Kasner had been trying to think of how to show children the difference between infinity and a number so large that we can barely imagine it. He needed a catchy name for that unthinkably large number, a name, in contrast to the number itself, that would be easy to think about. On a walk with Milton and Milton's brother, he asked if either of them had any ideas, and Milton came up with *googol.*

> You, too, can make up your own very large numbers and give them strange names. Try it. It has a certain charm, especially if you happen to be nine.
>
> **CARL SAGAN,** *COSMOS*

At this late date—Milton died in 1980—we can only speculate about why that was the word he thought of. Did he like the burbly, messy, baby sound of *goo*? Or did it sound monstrous, like a name for an ogre? Had he heard of Barney Google* with the goo-goo-googly eyes?

Back when Milton thought of it, there wasn't much use for the numeral googol. It was a speculative exercise. But now,

***Barney Google* was a comic strip originally drawn by Billy DeBeck and first published in 1919. It was later called *Barney Google and Snuffy Smith,* and later just *Snuffy Smith.* DeBeck contributed to the American vernacular such words and phrases as sweet mama, horsefeathers, hotsy-totsy, and heebie-jeebies. The goo-goo-googly eyes song had lyrics by showman Billy Rose. It was a big hit in the 1920s.

with fast algorithms and faster computers, working with googol—for mathematicians at least—is somewhat routine. Do you need to think about the number of subatomic particles in the visible universe? Work with googol.

Googol, let it be noticed, isn't spelled the same way as the popular Internet search engine Google. That may have been an accident. The legend is that the founders of Google meant to use Googol, but along the way, somebody simply spelled it wrong. They do sound pretty much the same. Another version of the story is that googol.com wasn't available, but google.com was; as simple as that. Some say that Google was initially a misspelling on an investor's check, and rather than complicate things, the company was named to match the check. The 2006 Google home page has it that "Google's use of the term [googol] reflects the company's mission to organize the immense, seemingly infinite amount of information available on the web."

Key word in that sentence is really *seemingly*. Googol is not infinite. No number is. Infinity is not a number—you can't count it, divide by it, multiply by it, or add or subtract it. Infinity is a concept. If you could number it—if it were a number—it wouldn't be infinite. No matter how large a number is, if you add one more, it becomes larger—and that's the way to reach infinity, though, of course, infinity can't be reached.

There are—and as long as we can add one more, there have to be!—numbers larger than googol. A googol is ten to the one hundredth power. A centillion is ten to the three hundred and third power. That almost makes a googol look small.

> **There is a fifth dimension beyond those known to man. It is a dimension vast as space and timeless as infinity. It is the middle ground between light and shadow, between the pit of his fears and the summit of his knowledge. This is the dimension of imagination. It is an area called the Twilight Zone.**
>
> ROD SERLING, *THE TWILIGHT ZONE*, TV SERIES, 1959–1964

We're not getting to the end of numbers (you know that there's no end, right?)—but we are getting to the end of numbers with a name. Back at Google, the home offices are called Googleplex, and this time the misspelling was deliberate. A *googolplex* is the next new number word after googol. Milton Sirotta also invented googolplex. As I write, googolplex is the last number to be given a name.

> Equations are more important to me [than politics], because politics is for the present, but an equation is something for eternity.
>
> ALBERT EINSTEIN

How big is a googolplex? Well, Mr. Bones, it's so big that . . . It's so big that it's one with a googol zeros after it. It's so big that it's ten raised to the power of a googol, ten to the tenth power to the hundredth power.

It's so big that it's more than enough to count the length of the days, weeks, and months, the seasons and the years, since creation, since the birth of the universe, since the beginning.

It's so big that if you went to the farthest star in the farthest galaxy of the farthest corner of the universe, and while you were getting there (in your space suit and your rocket ship and with the elixir that allowed you to live forever) you left a trail behind you every single *inch* of the way, and the trail was of zeros, one after another, sprinkled like bread crumbs through the stars and the galaxies and the coldness of space, you would run out of room before you ran out of zeros.

And this is so, this impossible trail of zeros, despite the simple fact that a googolplex is a finite number. When you'd finished your googolplex of zeros, when you'd gone beyond the edge of the universe, trailing behind you that string of zeros so that you could find your way home again, you would be no closer to infinity than when you started. Infinity would remain

> A googol is exactly as far from infinity as the number one.
>
> CARL SAGAN

just as far away as it has always been, whether you measure from one or from googolplex.

But still, when you were there, at the end of all your zeros, zero after zero after zero, you could still add one more—to make googolplex plus one—to the line of numbers that you drew through the stars, connecting them zero by zero, to home, to one, to nothing, and to all.

# BIBLIOGRAPHY


Crump, Thomas. *The Anthropology of Numbers.* Cambridge University Press, Cambridge. 1992.

Dantzig, Tobias. *Number: The Language of Science.* Macmillan, New York. 1967.

Humez, Alexander, Nicholas Humez, and Joseph Maguire. *Zero to Lazy Eight: The Romance of Numbers.* Simon & Schuster, New York. 1993.

Ifrah, Georges. *From One to Zero, a Universal History of Numbers.* Viking, New York. 1985.

——— *The Universal History of Numbers: From Pre-History to the Invention of the Computer.* John Wiley & Sons, New York. 2000.

Kaplan, Robert. *The Nothing That Is: A Natural History of Zero.* Oxford University Press, New York. 2000.

Livio, Mario. *The Golden Ratio: The Story of Phi, the World's Most Astonishing Number.* Broadway Books, New York. 2003.

Menninger, Karl. *Number Words and Number Symbols: A Cultural History of Numbers.* The MIT Press, Cambridge, MA. 1970.

Schimmel, Annemarie. *The Mystery of Numbers.* Oxford University Press, New York. 1993.

Smith, David Eugene, and Jekuthiel Ginsburg. *Numbers and Numerals.* National Council of Teachers of Mathematics, Washington, D.C. 1961.

Wells, David. *The Penguin Dictionary of Curious and Interesting Numbers.* Penguin Books, London. 1997.

# INDEX